Better Homes and Gardens®

STEP-BY-STEP BASIC CARPENTRY

Step-By-Step Basic Carpentry

Editors: Larry Clayton, Jim Harrold
Copy and Production Editor: David A. Walsh
Graphic Designer: Thomas Wegner
Contributing Writers: Greg Erickson,
Gary Havens, David Kirchner
Technical Consultants: George Granseth,
Al Roux, Don Wipperman
Drawings: Carson Ode

Large-Format Edition. Fifth Printing, 1988.
Library of Congress Catalog Card Number: 80-68452
ISBN: 0-696-01185-9

CONTENTS

INTRODUCTION

Ever watch a magician pull a rabbit from a hat? If so, you know that there's more to the trick than a few magic words and a wave of the wand. That "more" is a knowledge of the *basics* of illusion, the *right equipment,* and *lots of practice.*

These same elements can make you into a pretty fair home carpenter, too. That's because nothing is intrinsically difficult about carpentry, no matter whether you're building a birdhouse or a room addition. With a knowledge of the basics of carpentry, the right equipment, and practice, you should be able to take on almost any project and do it well.

That's not to say, however, that after digesting the material in this book, you'll be able to conquer every carpentry challenge. You won't. What the book will do is provide you with a good foundation on which you can build your carpentry expertise.

Because getting off to a good start is important with any hobby, we begin with a section titled "Start Out Right." In it we talk about the importance of setting up shop and how to do it, the basic tools you'll need, and the many materials and hardware items you'll encounter as you begin your carpentry career.

Then it's on to the real tricks all carpenters use to pull off their magic acts—*techniques.* In "Master These Basic Techniques," you'll learn how to measure, cut, drill, fasten, shape, smooth, and finish—the techniques common to all carpentry projects. In addition, a short section tells how to keep your tools razor-sharp.

So that you can put your newfound knowledge to work, we've included several projects in the section "Try These Basic Projects." In it you not only will come across the basic box (and several things you can do with it), but also find out how to hang shelves on walls, and how to frame or furr-out walls as well as how to install drywall and sheet paneling. And you'll be relieved to know that you can complete all of the projects featured here without having to resort to using expensive power tools or any tricky joinery techniques.

Likewise, none of the projects requires that you have a permit from any municipality or an after-completion inspection by building officials. If you decide to take on other, larger-scale projects such as adding to an existing structure or making other structural changes, check with local officials as to their requirements.

A word about safety: As with any other do-it-yourself activity, THINK SAFETY is the watchword with carpentry. Be especially watchful of how you use the carpentry tools whose purpose it is to cut. They don't discriminate. For more information on this important topic, please turn to and carefully read page 13.

START OUT RIGHT

Why spend 20 pages of a 96-page book talking about how to get your carpentry career off to a good start? Because it's the best way we could think of to introduce you to this wide, wide world you're about to enter.

In it you'll discover a myriad of products and materials. You'll also encounter some pros and sales help who assume you are familiar with the things they deal in each day. The only way to keep confusion to a minimum is to be prepared by reading this section.

Because one of your first concerns probably will be where to do your carpentry projects, that's where this chapter begins. As you'll see, your shop can be in any number of places.

Then we discuss the tools you'll need—both power and hand tools. You may be surprised at how few you actually need to complete most projects.

After this, you'll learn about how to select and handle the many materials you'll come in contact with, as well as the potpourri of hardware items you'll see at most building materials outlets.

Setting Up Your Home Workshop

No matter what the hobby a person involves himself with, having an area set aside especially for that pursuit makes good sense. And carpentry is certainly no exception. You need a convenient, comfortable, well-organized place to work and to store all the tools and other paraphernalia you'll accumulate as you continue doing various carpentry projects.

Your carpentry headquarters can be in any number of places around your home, can be large or small, and can be basic or elaborate. But have one you must—and the sooner the better.

Locating Your Work Center

Where you locate your workshop depends on available space and your needs. Basements, garages, seldom-used rooms, even closets and attics are all candidates. But if you have a basement, look there first, as it has several distinct advantages over any other area. It's off the beaten path, so you needn't worry too much about disrupting family activities as you work. In most homes it's also one of the few areas with sizable amounts of unused space—an important factor especially if you are eventually planning to equip the shop with stationary power tools. And, basements offer a comfortable environment to work in—moderately warm in winter, pleasantly cool in summer.

Not all basements will work, however. Access and dampness can be problems. Obviously, if you can't get materials in and out, or if you can't control dampness by waterproofing the walls or dehumidifying the area, you'll have to look elsewhere.

Planning Pointers

Once you've decided on your shop's location, you need to spend some time planning how to equip it. The workshop shown below illustrates many of the items you'll want to include in yours, space permitting. Also keep the following pointers in mind as you plan.

• Because a workbench is the activity hub of every workshop, selecting one that's right for your situation is an important first consideration. Full-size workbenches typically measure 6 to 8 feet long, 24 to 36 inches deep, and 40 to 42 inches high. Obviously, however, not all shops can contain a bench of these dimensions. Likewise, not all people will find the heights listed comfortable. So you may have to do some tailor-*(continued)*

Planning Pointers *(contd.)*

ing. To determine the right work surface height for you, for example, measure from the floor to your hipbone.

Whether you buy or build the bench is up to you. Commercially made ones range from simple steel and particleboard arrangements to elaborate versions that would satisfy even the most dedicated cabinetmaker.

If you elect to build your own, use building materials and fasteners that will yield a strong, stable work surface. We've included a couple of well-designed benches here—one full-size one, and a closet-size version—both of which can be knocked together in a weekend or less.

- Be sure to provide for plenty of storage, both for your tools and for the many containers and other supplies that otherwise will quickly clutter up the area. Use perforated hardboard racks to store small hand tools within easy reach, and get yourself an inexpensive organizer for keeping all those nails, bolts, screws, and other hardware items in their place. Store power tools and flammable liquids in locked cabinets, if possible, or another *safe* place.
- Arrange for adequate shop lighting. You'll want one or more lights for general illumination as well as more intense task lighting for those workshop operations that demand it.
- Run at least one 20-amp electrical circuit with ground fault circuit protection to the shop to provide power for your power tools and lighting needs. Large shops should have separate circuits for tools and lights. Position electrical outlets strategically around the workshop so that power is never far away.
- To ensure adequate ventilation, install an exhaust fan capable of

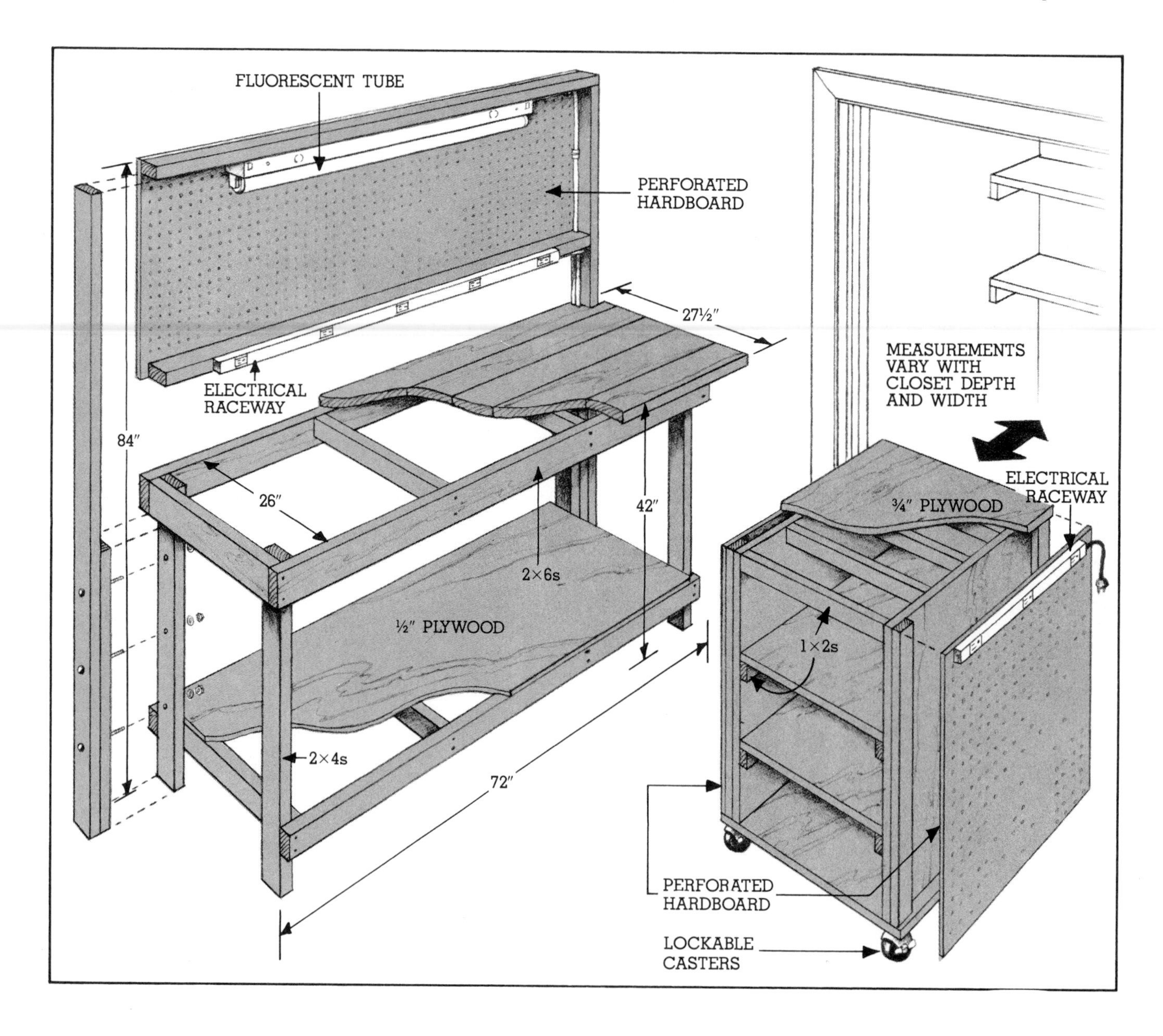

changing the air in the shop every four minutes. The number of cubic feet (length x width x height) in your shop determines the size fan needed.

- To warn against fire, you'll want a smoke detector as well as a good-size ABC-rated fire extinguisher. Also, have a first-aid kit handy to deal with injuries if they occur.
- To facilitate cleanup, have a broom and dustpan or a shop vacuum on hand. And don't forget to get yourself a metal trash container with a tight-fitting lid to safely house the many wastes that will result from your carpentry projects.
- To support bulky sheet goods and lengths of lumber while they're being measured and cut, you'll need a couple of sawhorses. We show three different possibilities here. If space is at a premium in your shop, consider buying the hinged metal leg horses or the metal bracket type, both of which you can break down for easy storage. Otherwise, you may prefer to build your own entirely from wood.
- You'll want to have some way to conveniently carry your tools from place to place as you work. The tool caddy shown here has plenty of room for many tools at one time. It's deep enough to hold a few power tools, and long enough to accommodate a crosscut saw or ripsaw and a carpenter's level. Holes drilled through the shelf provide grab-it-quick storage for a hammer, screwdrivers, and other hand tools.

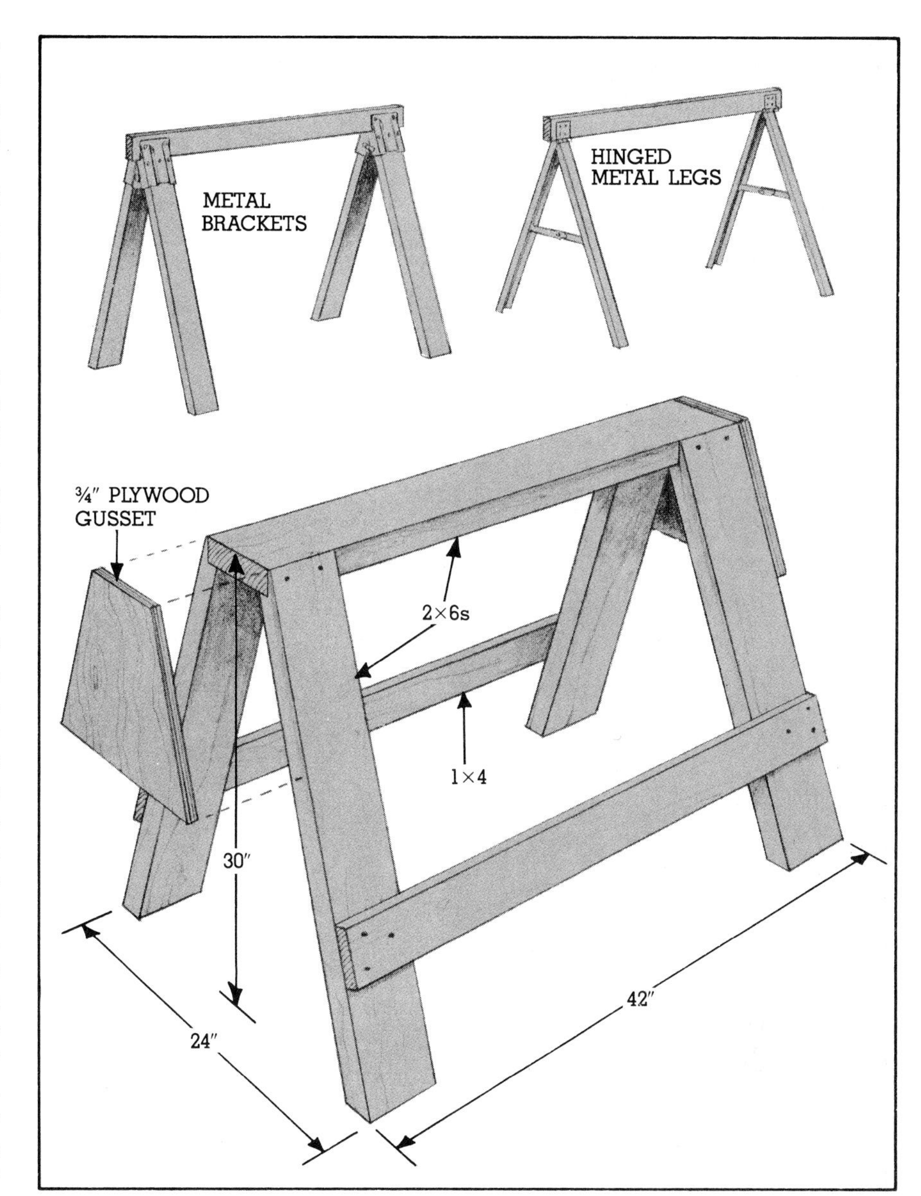

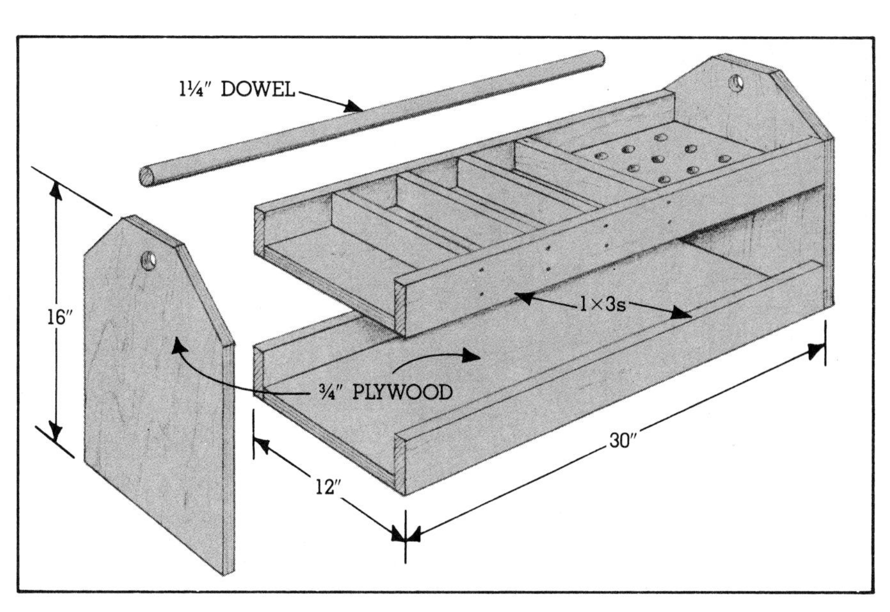

Buy the Basic Tools

For many home carpenters, purchasing tools is a never-ending process. A tool exists for every conceivable need. But as a beginning carpenter, you don't need all the "extras." Instead, concentrate on assembling a few key tools—those shown and discussed here and on the following three pages—then add to this core group others as the need arises. (Note: When tool shopping, buy the best you can afford. Good-quality tools, properly used and cared for, often last a lifetime.)

Hand Tools

1 Few if any carpentry tools see more action than the *flexible steel tape,* so make it one of your first purchases. A 12-footer with a lock-button and plastic coating on the tape itself offers adequate measuring flexibility.
2 Use a *framing square* to square almost anything, to check stud spacings quickly, even to mark rafter and stair stringer cuts.
3 Square up smaller pieces and mark 45-degree angles with a *combination square.* Look for one with a built-in level and scriber.
4 Duplicate angles of from 0 to 180 degrees with a *T bevel.*
5 Plumb and level large projects with a *carpenter's level.* Buy a 24- or 28-inch model.
6 Mark cutoff lines and make pilot holes with an *awl.*
7 Snap long, straight lines with a *chalk line reel.* This tool also doubles as a plumb bob to establish true vertical lines.
8 A 26-inch, 8-point crosscut *handsaw* will handle most general-purpose cutting chores.
9 To make perfect miter joints, you need a *backsaw* and *miter box.*
10 And to cut curves and straight lines easily in tight places, use a *keyhole saw* with its narrow, tapered blade.
11 Cut intricate curves in thin materials with a *coping saw.*
12 *Wood chisels* enable you to shape mortises, cut and smooth wood joints, and do other specialized cutting work. Look for ones with metal-capped handles.
13 And don't be without a retractable-blade *utility knife* for cutting chores that require a razor-sharp blade.
14 For driving and removing nails and other fasteners, buy a 16-ounce *curved-claw hammer.*
15 And to sink the heads of finishing nails below the surface of the work, use a *nail set.* Purchase several sizes.
16, 17 You'll need a set of *slotted* and *Phillips-tipped screwdrivers* or a *ratchet-action screwdriver* with several tips for driving and removing screws.
18 To fasten nuts, bolts, and lag screws, buy a pair of 10-inch *adjustable-end wrenches.*
19 If you don't have a pair of *slip-joint pliers,* add them to your tool collection right away.
20 For holding about-to-be-sawed, -drilled, or -joined members, invest in several *C clamps* of varying sizes. Later, you may want to add pipe and miter clamps, too.
21 For good results when planing with the grain, purchase a *jack plane.*
22 A *block plane's* specialty is planing across grain.
23 The triangular teeth of a *rasp* remove wood in a hurry. Look also at the rasp-like forming tools available. These have replaceable blades.
24 *Wood files* smooth edges and do other light smoothing duty. The best purchase here is a coarse, half-round, double-cut file.
25 And purchase an *oilstone* for various tool-sharpening needs.

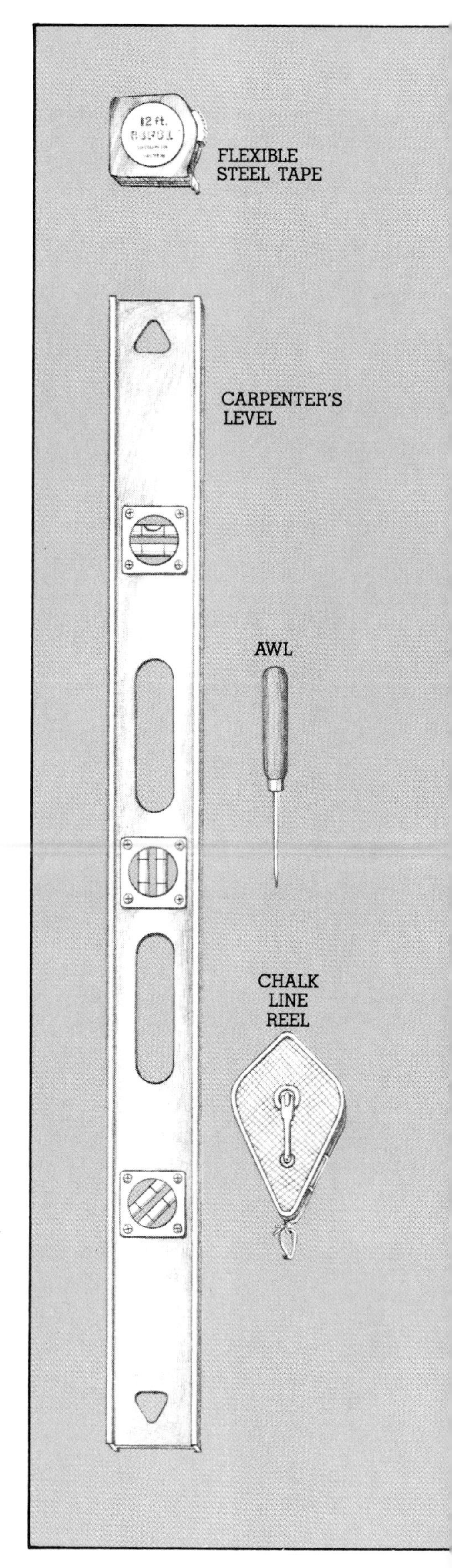

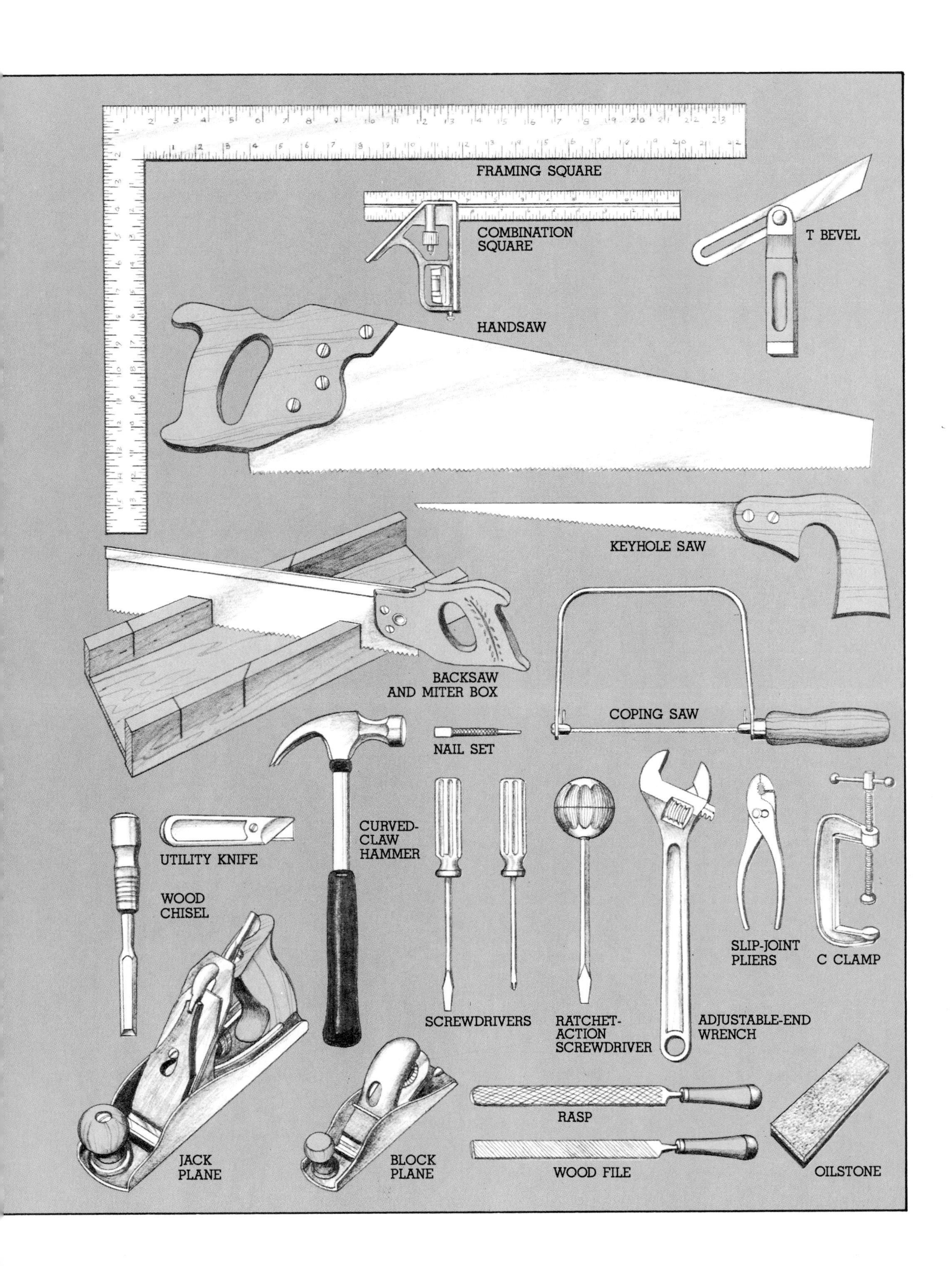
FRAMING SQUARE
COMBINATION SQUARE
T BEVEL
HANDSAW
KEYHOLE SAW
BACKSAW AND MITER BOX
COPING SAW
NAIL SET
CURVED-CLAW HAMMER
UTILITY KNIFE
WOOD CHISEL
SCREWDRIVERS
RATCHET-ACTION SCREWDRIVER
ADJUSTABLE-END WRENCH
SLIP-JOINT PLIERS
C CLAMP
RASP
WOOD FILE
OILSTONE
JACK PLANE
BLOCK PLANE

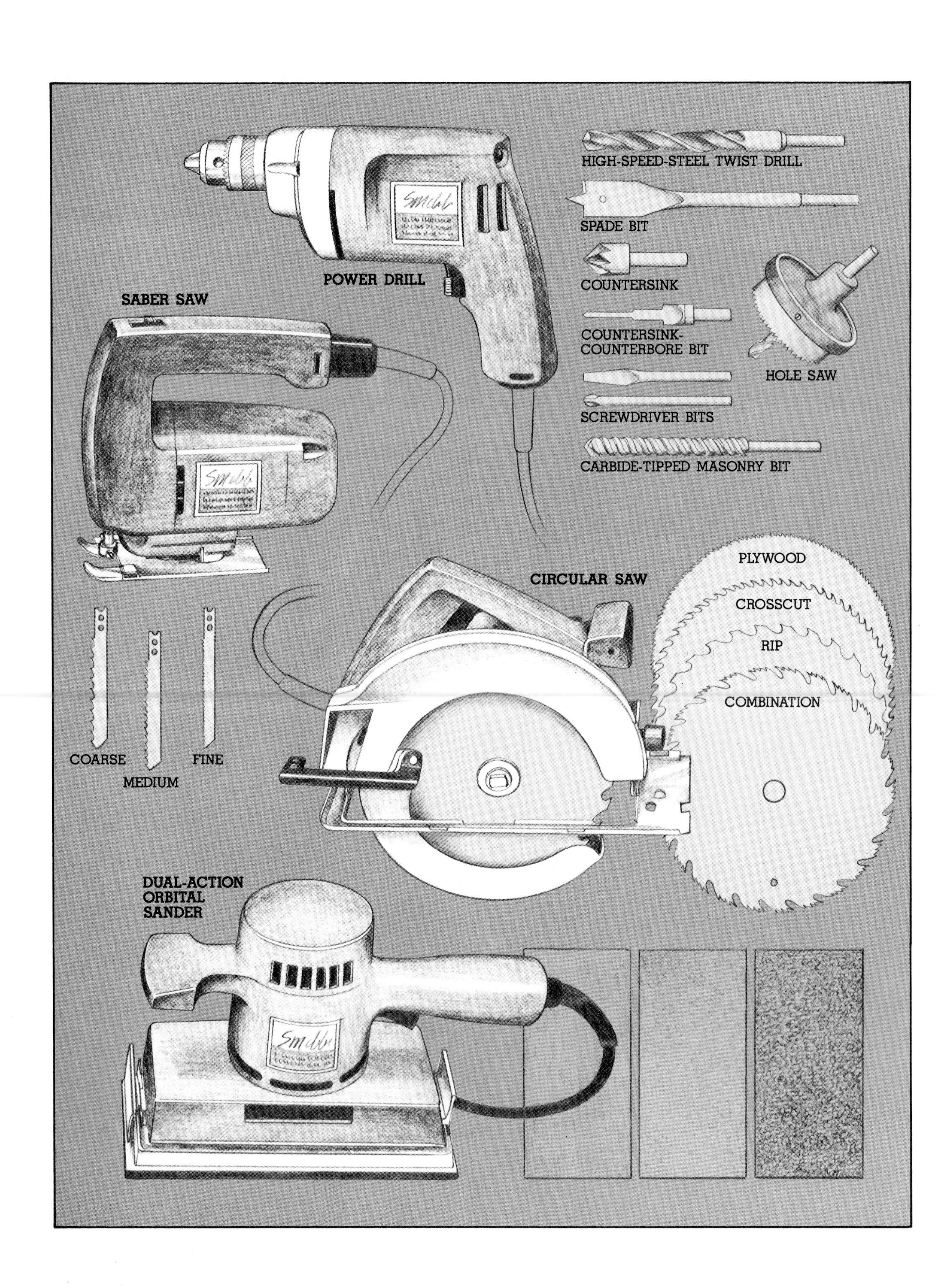

POWER DRILL
HIGH-SPEED-STEEL TWIST DRILL
SPADE BIT
COUNTERSINK
COUNTERSINK-COUNTERBORE BIT
HOLE SAW
SCREWDRIVER BITS
CARBIDE-TIPPED MASONRY BIT
SABER SAW
CIRCULAR SAW
PLYWOOD
CROSSCUT
RIP
COMBINATION
COARSE
MEDIUM
FINE
DUAL-ACTION ORBITAL SANDER

Affordable Power Tools

You may not associate power tools with the basic tool kit; but if you think in terms of the convenience and speed that power drills, saber saws, circular saws, and dual-action orbital sanders bring to the home workshop, you really can't afford to be without them. Following is a rundown on each member of this power tool quartet.

Power drills come in three sizes—¼ inch, ⅜ inch, and ½ inch. These fractions describe the maximum diameter of bit shafts the drills can accept. The ⅜-inch model, because it's sturdy enough to handle all but the heaviest types of work, is your best bet. Make sure the one you choose has a reverse switch and a variable-speed trigger. Both features come in handy if you use your drill as a screwdriver.

As you can see, a power drill can drive a variety of bits. With *high-speed-steel twist drills* you can bore up to ½-inch holes in wood or metal. To make ¼- to 1½-inch-diameter holes in wood, get a set of *spade bits.* And if you have a need for even larger holes—up to 2½ inches—purchase a *hole-saw* set or an *adjustable hole saw. Carbide-tipped masonry bits,* not surprisingly, find their way through masonry and concrete surfaces. *Countersinks* and *countersink-counterbore bits* make form-fitting holes for screws. And *screwdriver bits* will speedily transform your reversible drill into a power screwdriver.

No tool in the carpenter's workshop can claim the versatility of the portable *saber saw.* This little giant can crosscut, rip, angle cut, bevel, even cut holes and scrolls in almost any material. Just put in the correct blade for the job, and you're off and running.

You can select from dozens of blades to cut through a variety of materials. As a general rule, though, the fewer number of teeth the blade has, the faster and rougher the cut will be. The more teeth per inch, the slower and smoother the result.

Circular saws can make many of the same cuts its saber saw counterpart can, only faster.

When shopping for one of these, look for a 7¼-inch model with an automatic blade guard, a saw blade depth adjustment, an adjustable baseplate so you can make bevels, and a ripping fence accessory.

The saw you buy probably will come with a *combination blade,* which means it's designed for both across-the-grain and with-the-grain cuts. You may later want to buy a different blade for each of these two types of cuts—*crosscut* and *ripping.* And for making clean cuts in sheet goods, get yourself a *plywood* blade. You may want to invest in tungsten-carbide-tipped blades rather than less-expensive but short-lived standard blades.

A dual-action *orbital sander* makes quick work of almost any sanding job. In its orbital mode, it removes excess material fairly quickly. Then with the flip of a switch, the action changes to straight-line sanding, the perfect motion for finish sanding.

So you'll be prepared for any sanding eventuality, you should lay in a supply of coated abrasives. Your selection should include several sheets in varying degrees of coarseness—coarse, medium, and fine. (Note: The larger the grit number, the finer the texture of the abrasive. For example, 150-grit is *fine,* whereas 60 is *coarse.*) For most of your sanding needs, aluminum oxide abrasive is probably your best bet, all things considered.

Tool Safety Tips

Experienced home carpenters know that tool safety is a blend of exercising common sense and following certain guidelines when working in and around the shop. To make all of your experiences with tools happy ones, keep in mind the following:

- Use tools *only* for the jobs they were designed to do. If a tool came with an instruction manual, take the time to read it to find out what it will do and will not do.
- Check on the condition of a tool before using it. A dull cutting edge or a loose-fitting hammer handle, for example, spells trouble. Also inspect the cord of a power tool to make sure it's not frayed.
- Don't work with tools if you're tired, in a bad mood, or in a big hurry.
- Wear goggles whenever the operation you are performing could result in eye injury.
- The safety mechanisms on power tools are there for your protection. Do not tamper with or remove them from the tool.
- Don't wear loose-fitting clothes or dangly jewelry while you are using tools, especially power tools.
- Keep children and others at a safe distance while you're using any tool. And before letting children use a tool, instruct them on how to operate it.
- Before servicing or adjusting a power tool, unplug it and allow moving parts to come to a standstill.

Selecting and Handling Materials

Choosing and Buying Lumber

As a home carpenter, you will be buying lots of lumber for various projects. And although you needn't become a lumber "expert," smart buymanship requires that you know what's available, when to use what type and size lumber, and how to order this most basic of all building materials. After reading this and the next three pages, you'll know the basics.

Begin by realizing three things about lumber. First, there are only two basic types—*softwoods* and *hardwoods*. Second, keep in mind that both types have a set of *nominal* dimensions (what you order) and a set of *actual* dimensions (what you get after the lumber is milled and dried). And third, remember that both softwoods and hardwoods have different grading systems, which we explain on pages 16–17.

Now, take a look at the chart, opposite. It classifies lumber into five groupings, then lists common uses for each, as well as nominal and actual dimensions of various-size members.

Because of the array of lumber thicknesses and widths, lumberyard and home center personnel use the *board foot* to measure and price the amount of wood in a given piece. (As shown below at left, a board foot is defined as the wood equivalent of a piece 12 inches square and 1 inch thick.) For most of your needs, though, just specify a piece's length in linear feet and the yard will compute the cost.

When lumber shopping, come prepared with a list detailing your needs. To order, state the quantity, thickness, width, length, grade and specie, for example—four 2" × 4" × 8' No. 2 fir. And after placing your order at the desk, walk out into the yard to see firsthand what you're getting. If you spot any of the defects shown below at right and if they'll seriously interfere with how you intend to use the lumber, ask for replacements. And to prevent some of these problems from occurring once the goods are in your hands, follow the handling and storage advice given on page 21.

(continued)

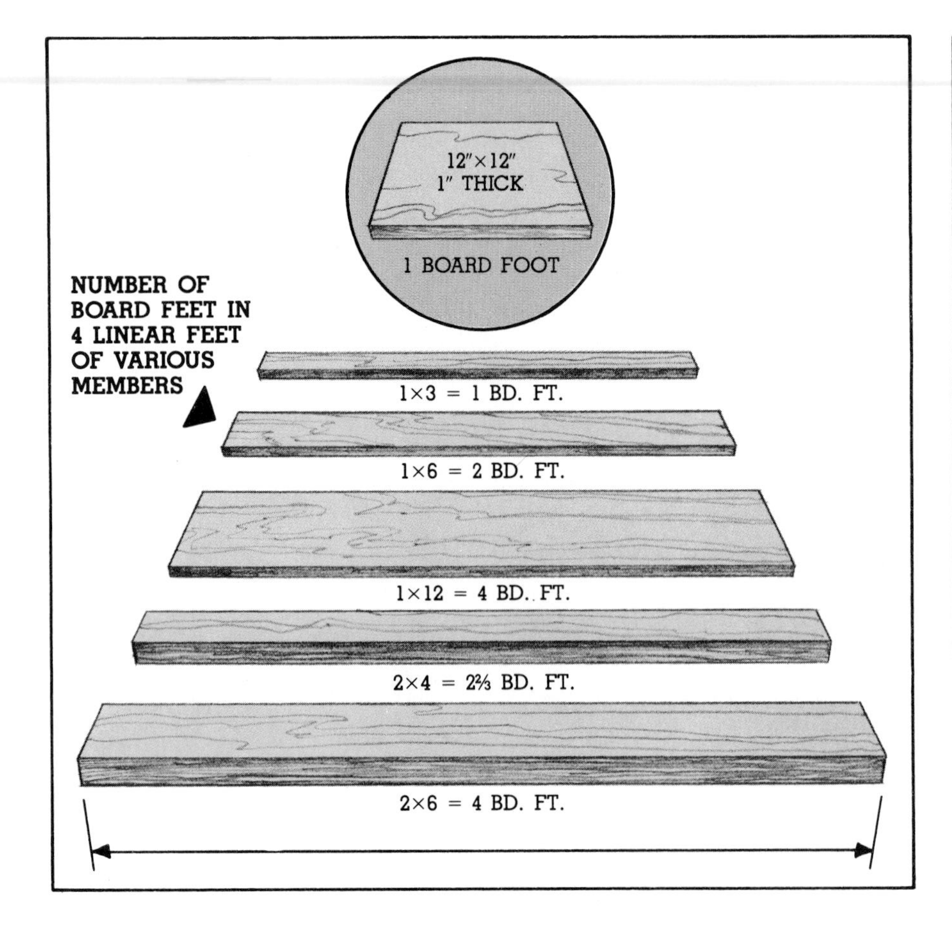

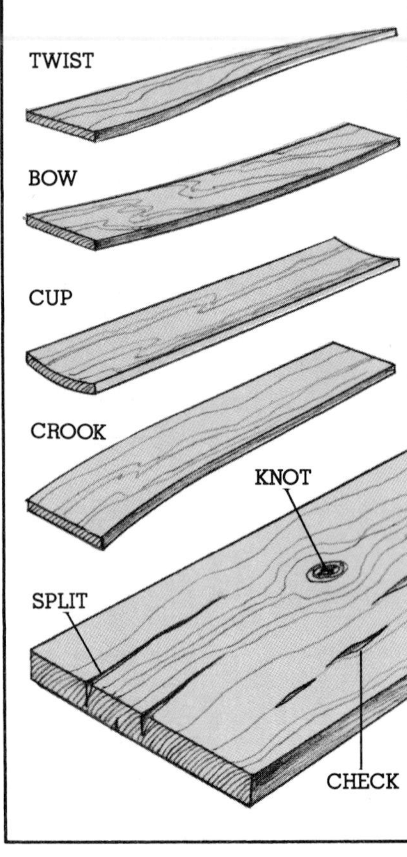

Lumber Selector

Type	Common Uses	Nominal Sizes	Actual Sizes
Strips	Furring for wall-paneling material (drywall, hardboard, plywood) and ceiling material (drywall, composition tiles); trim; shims; spacers; blocking; bridging; stakes; forms; crates; battens; light-duty frames; edging; latticework.	1 × 2 1 × 3	¾ × 1½ ¾ × 2½
Finish Lumber: Boards	Interior paneling; exterior sheathing; structural framing and finishing; exterior siding and soffits; subflooring and flooring; decking; fencing; walks; interior and exterior trim; fascias; casing; valances; shelving; cabinets; closet lining; furniture; built-ins.	1 × 4 1 × 6 1 × 8 1 × 10 1 × 12	¾ × 3½ ¾ × 5½ ¾ × 7¼ ¾ × 9¼ ¾ × 11¼
Tongue and Groove	Subflooring; flooring; exterior sheathing; exterior siding; decorative interior wall treatments.	1 × 4 1 × 6 1 × 8 1 × 10 1 × 12	Actual sizes vary from mill to mill.
Shiplap	Exterior sheathing and siding, decking; underlayment; subflooring; roof sheathing; decorative interior wall treatments.	1 × 4 1 × 6 1 × 8	¾ × 3⅛ ¾ × 5⅛ ¾ × 6⅞
Dimension Lumber	Structural framing (wall studs, ceiling and floor joists, rafters, headers, top and sole plates); structural finishing; forming; exterior decking and fencing; walks; benches; screeds; stair components (stringers, steps); boxed columns.	2 × 2 2 × 3 2 × 4 2 × 6 2 × 8 2 × 10 2 × 12	1½ × 1½ 1½ × 2½ 1½ × 3½ 1½ × 5½ 1½ × 7¼ 1½ × 9¼ 1½ × 11¼
Posts	Heavy-duty structural framing; support columns; fencing; decking; turning material for wood lathes; building material for architectural and decorative interest.	4 × 4 6 × 6	3½ × 3½ 5½ × 5½
Timbers	Heavy-duty structural framing; support columns; building material for architectural and decorative interest.	Rough-sawn; sizes vary.	Actual sizes vary slightly up or down from nominal sizes.

Choosing Softwoods

At this point, you may not know one softwood from another. (The chart below compares the most commonly available species.) But don't let that bother you a great deal because unless you're framing a major structural component that will bear great weight, such as a ceiling joist, you can't make a serious mistake when buying softwoods.

You can also take comfort in knowing that many building material home centers and lumberyards, because they bulk-buy lumber to achieve greater savings, stock only a few species. As a result, you may find you can actually do very little selecting in most cases. You may have only one or two species to choose from. However, you will be asked what grade lumber you want, which in large part depends on the nature of your project.

Admittedly, softwood lumber grading is tricky at first. That's partially because several grading systems exist. But in general, you'll be on safe footing if you think in terms of two overall grade classifications: *select* and *common.* Use select lumber *(B and Better, C,* and *D)* for showy projects such as cabinetry where good appearance is a vital consideration. For all other projects, common lumber *(Nos. 1, 2, 3,* and *4)* will do nicely. Not surprisingly, the better the grade—that is, the more defect-free—the more you will pay.

Talk over your project with sales personnel, too; often, they can suggest the best buy for your particular need.

Softwood Selector

Species	Characteristics	Common Uses	Additional Information
Cedar, Cypress	Lightweight but good flex strength; easy to work with; take nails well; highly resistant to rot; no preservative needed for exterior use; cedar has a sweet aroma.	Trim; paneling; decks; fencing; posts; chests; closet lining (aromatic cedar); shingles (cedar).	Cedar 1×4s, 1×6s, and 1×8s usually come with one smooth-cut face, one rough-cut face. Cedar is commonly available in the North; cypress, in the South.
Fir, Spruce, Pine, Hemlock	Easy to work with; finish well; good strength; weights vary; resist shrinking.	House framing; paneling; trim; interior furniture; decking; fencing; sheathing; general utility.	Spruce also is known as "white wood." You can use these species almost interchangeably in small projects. Apply a preservative if used outdoors. All will split if carelessly nailed near ends of boards; drill pilot holes or blunt nail ends to prevent this. This classification includes Southern (Yellow) Pine, which is very yellow in color, strong, but also brittle. Nails may not "seat" well in this wood.
Treated Lumber	*Pressure treated:* Heavy and strong; usually greenish in color; tends to shrink; can be painted, stained, or left to weather naturally; highly resistant to rot. *Creosote-* or *penta-treated:* Solution applied only to surface; not as rot-resistant.	Outdoor projects of any kind: soffits, fascias, decking, fencing, posts.	Use gloves to protect your hands against splinters, which sting and burn. Never use scraps from this chemically laden wood in your fireplace or wood-burning stove.
Redwood	Lightweight; fairly strong but brittle; can snap under enough pressure; highly resistant to decay; no preservative needed; cuts cleanly and easily, but splits easily, too; weathers quickly.	Trim; paneling; decks; fencing; fascias; outdoor furniture; house siding.	For posts and near-ground structural members, *heart-wood* is best. For all other projects, save money by using *garden-grade* redwood. Drill pilot holes before nailing to avoid splitting.

Choosing Hardwoods

You can readily buy all kinds of papers, vinyls, and laminated plastics that *look* like hardwood, but nothing quite measures up to the real thing.

Unfortunately, genuine hardwoods are an increasingly scarce commodity. The demand for their unique aesthetic and structural qualities—some of which you'll find listed in the chart below—far outstrips the amount of lumber produced by these slow-growing deciduous trees. The result: You pay a premium price for most hardwood lumber or settle for a man-made facsimile.

Hardwood lumber, unlike softwoods, is milled to make use of virtually every splinter. So instead of the standard sizes softwoods come in, hardwoods are sold in pieces of varying lengths and widths, usually from ½ to 2 inches thick (nominal size). Individual boards generally are smooth-surfaced on two sides (S2S), and are priced at so much per board foot (see page 14).

Hardwood grading differs from softwoods, too—it's based primarily on the amount of clear surface area on the board. Heading the list is *FAS Grade* (Firsts and Seconds), which is the most knot-free, followed by *Select Grade, No. 1 Common,* and *No. 2 Common.*

Most lumberyards can't afford to maintain an extensive inventory of hardwood lumber, and generally stock only a limited assortment of a few species such as birch, mahogony, or oak. For the best selection, try to find a yard that specializes in hardwoods. You'll be astounded at the exoctic species they stock or can special order.

Hardwood Selector

Species	Characteristics	Common Uses	Additional Information
Ash	Heavy; strong; hard; rigid; works well with tools; holds nails and screws well but is prone to splits.	Formed parts of furniture; cabinets; millwork; baseball bats; implement handles.	Ash has a grain pattern similar to oak, but wilder and more yellow.
Cherry	Heavy; strong; hard; rigid; difficult to work with hand tools; resistant to shrinking and warping; fine grained; sands very smooth.	Fine furniture; cabinets; gun stocks.	Cherry is also known as fruitwood. Because of its scarcity, it is expensive.
Mahogany	Durable; easily worked; resistant to shrinking, warping, and swelling; fine grained; finishes well.	Fine furniture; cabinets; millwork; moldings; plywood veneers.	Genuine (Honduras) mahogany has a more pronounced grain than Philippine (lauan) mahogany.
Maple and Birch	Heavy; strong; hard; rigid; difficult to work with hand tools; resistant to shrinking and warping; fine grained; finish well.	Furniture; cabinets; millwork; moldings; flooring; inlays; veneers; butcher blocks.	Maple and birch have similar characteristics, although birch is somewhat softer and less expensive.
Oak	Heavy; strong; hard; rigid; difficult to work with hand tools; resistant to swelling; open grained; finishes well.	Fine furniture; cabinets; millwork; interior trim; flooring; stair rails.	Red oak is coarse grained and has a pink cast, and white oak is more decay-resistant and has a yellow cast.
Poplar	Lightweight and soft for a hardwood; quite easy to work; fine grained; paints well.	Furniture; cabinets; trim.	Poplar is gray-white in color with slight green streaks. It has some tendency to fuzz when sanded.
Walnut	Strong; hard; durable; works well with either hand or power tools; resistant to warping, shrinking, and swelling; fine grained.	Fine furniture; cabinets; millwork; paneling veneers; gun stocks.	Tropical walnut has a coarser, more porous grain than native North American walnut.

How To Select Sheet Goods

As you might guess by looking at the sketch on the opposite page, sheet goods can take many forms. Actually this is only a small sampling. But they all share certain characteristics that every do-it-yourselfer can easily appreciate. Quite simply, they're strong, easy to work with, and widely available.

What are *sheet goods*, or more correctly, *composition goods?* To produce *plywood*, manufacturers laminate thin layers (or plies) of wood to each other, using water-resistant glues for most *interior* types and waterproof adhesives for *exterior* plywoods. The front and back surface plies may be either softwood, usually Douglas fir, or hardwood.

Wood particles, sawdust, and glue, compressed and bonded together by heat, combine to form *particleboard*, *hardboard*, and other similar materials. And *drywall* is nothing more than gypsum powder sandwiched between layers of thin cardboard.

Smart buymanship begins with a knowledge of what's available. The chart below categorizes the sheet goods you'll find at most lumberyards and building material home centers, and discusses the typical uses of each.

Sheet Goods Selector

Material	Grades and Common Types	Thickness (in inches)	Common Panel Sizes (in feet)	Typical Uses
Plywood	Softwood plywood A-A; A-B; A-C; A-D	¼; ⅜; ½; ⅝; ¾	2x4; 4x4; 4x8	Projects in which appearence of one or both sides matters—cabinets, drawer fronts, bookcases, soffits, built-ins, shelves.
	C-D; CDX	⅜; ½; ¾;	4x8	Sheathing; subflooring; underlayment.
	MDO	⅜; ¾	4x8	Projects requiring an extra-smooth painting surface—tabletops, cabinets, outdoor signs.
	303 siding	⅜; ⅝; ¾	4x7; 4x8; 4x9	Exterior siding (smooth, grooved, reverse board-and-batten, and other textures); decorative wall and ceiling treatments.
	Hardwood plywood A-2 (good both sides)	¼; ¾	2x4; 4x8	Fine furniture and cabinetmaking; decorative wall panels.
	G1S (good one side)	¼	4x8	
Particle-board		⅜; ½; ⅝; ¾	2x4; 4x4; 4x8	Underlayment; core material for laminated furniture and counter tops; speaker enclosures; closet lining (aromatic cedar particleboard).
Hardboard	Standard; Tempered (moisture resistant)	⅛; ¼	2x4; 4x4; 4x8	Underlayment; drawer bottoms and partitions; cabinet backs; pegboard *(perforated hardboard)*; decorative wall panels (prefinished hardboard).
			4x8; 4x9 8", 9", and 12" widths; 16' length	Exterior siding in vertical and horizontal-lap formats. Available smooth and in shake, stucco, board-and-batten, and other textures.
Drywall	Regular; Moisture-resistant (MR); Fire-resistant (FR)	⅜; ½; ⅝	4x8; 4x12	Finish material for interior walls and ceilings; kitchen, bath, and basement surfaces (MR); surfaces in attached garages (FR); patch material for lath-and-plaster walls.

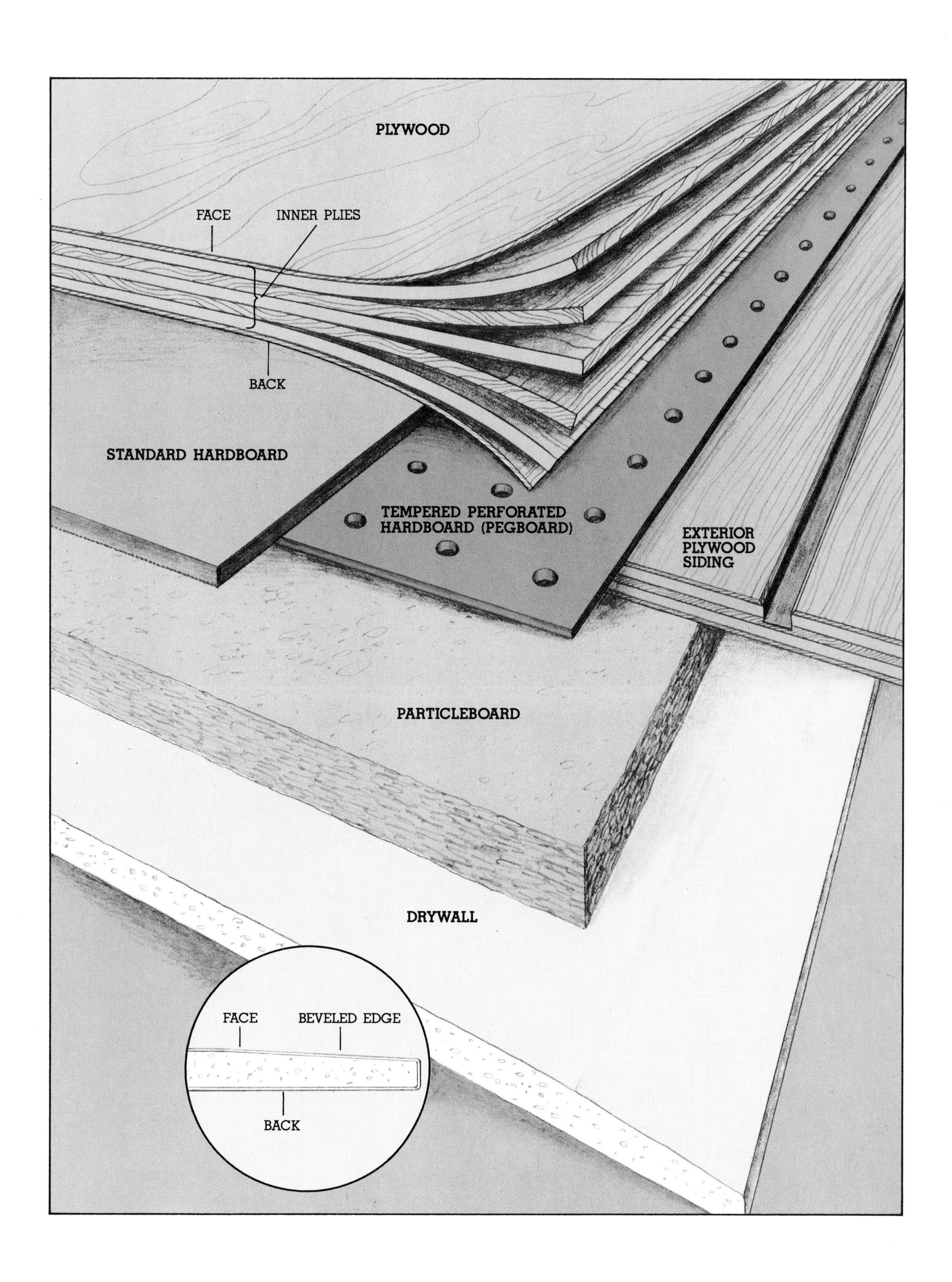
PLYWOOD
FACE
INNER PLIES
BACK
STANDARD HARDBOARD
TEMPERED PERFORATED HARDBOARD (PEGBOARD)
EXTERIOR PLYWOOD SIDING
PARTICLEBOARD
DRYWALL
FACE
BEVELED EDGE
BACK

Selecting and Ordering Moldings

If you could buy something that would add an instant decorative touch to your carpentry work, conceal unsightly seams and mistakes at the same time, and maybe even protect against damage from accidental nicks and bumps, would you?

Before you say that no such miracle material exists, take a look at the sketch and chart, at right. They show and discuss more than a dozen moldings, all of which can serve you well in one or more of the above-mentioned ways. You'll find all of these moldings and more at most lumber and millwork outlets, in styles that range from ornate and traditional to sleek and contemporary. And if you can't find a stock molding that suits your needs, some shops will mill shapes to your order. (This is often the only way to duplicate intricate moldings in older homes.)

Most moldings, which you can purchase in random lengths from 3 to 20 feet, are made of softwood, usually pine. A few of the most popular sizes and types, though, come in selected hardwoods—usually mahogany, oak, and birch. You can purchase softwood moldings unfinished or, if you're working with prefinished paneling or doors, prefinished to match.

To estimate your needs, make a list of each piece of molding and round each measurement up to the next largest foot. Doing this will ensure that you don't come up short of material.

When ordering, keep in mind that you can save money if you'll settle for random lengths purchased on a so-much-per-hundred-linear-feet basis rather than insisting on pieces of a specific length. And if you'll be painting the molding, you can save even more by ordering finger-jointed moldings—short pieces that have been joined end to end.

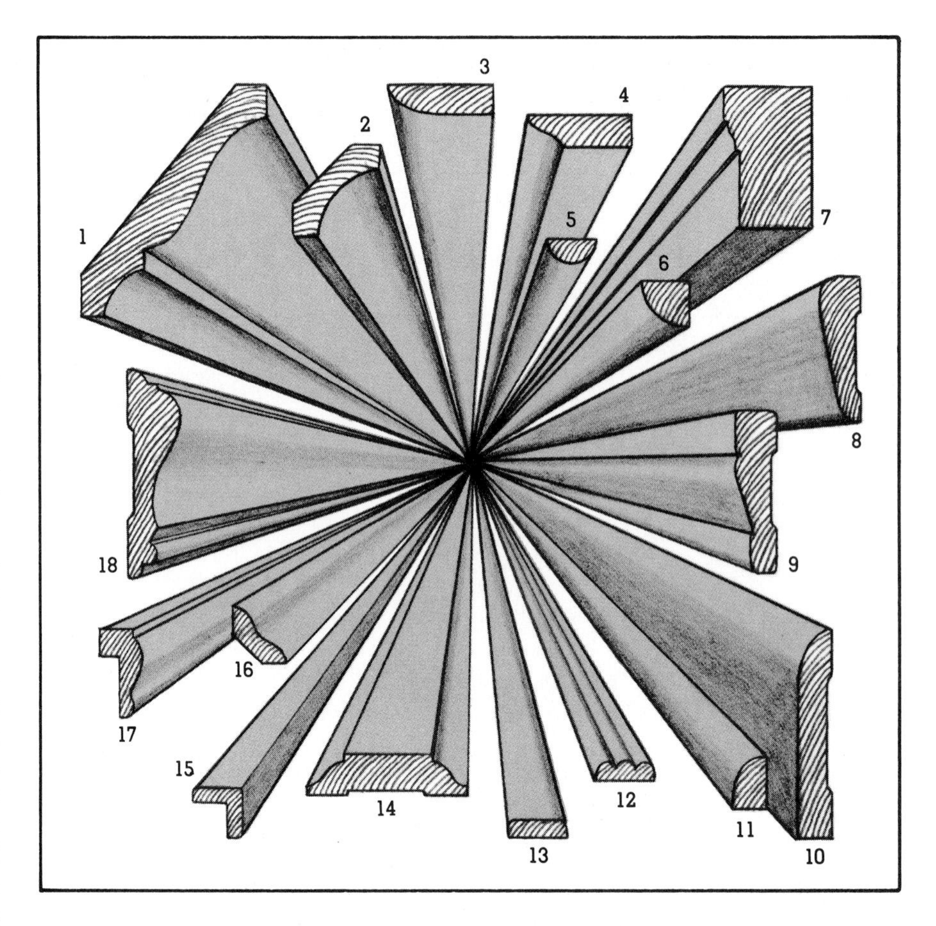

Molding Selector

Common Types	Typical Uses
1, 2 Crown, Cove	Trim and conceal joint between walls and ceilings.
3, 4 Stop	Attaches to faces of door jambs to limit door swing. Holds inside sash of double-hung windows in place.
5 Half-round	Serves as screen bead, shelf edging, and battens.
6 Quarterround	Serves as base shoe and inside corner guard.
7, 8, 9 Brick mold, Casing	Trim around openings for interior (casing) and exterior (brick mold) windows and doors.
10, 11 Baseboard, Base shoe	Trim and protect walls along wall-floor line.
12, 13 Screen bead	Covers seams where screen material fastens to frames. Finishes edges of shelves.
14 Batten	Conceals vertical or horizontal panel seams.
15, 16 Outside, Inside corner guard	Conceal seams and protect areas where walls or different wall finishes meet at corners.
17 Plycap	Conceals plywood edges. Caps top of wainscoting.
18 Chair rail	Protects walls from damage from chair backs. Hides seams where different wall materials meet.

Handling and Storing Materials

When you drive out through the exit gate of a building materials dealer with your materials, the assumption is that you're leaving with the correct amount of everything you ordered and that you're satisfied with its condition. Some suppliers even have you sign the purchase ticket to this effect. So from that moment on, keeping the materials in good condition becomes your responsibility. To keep from damaging your purchases, follow these tips as you handle and store your goods.

- To ensure a safe, uneventful trip home, make sure all materials are well secured to your vehicle with rope or baling twine. For large purchases, or if your car can't safely transport building materials, you can (for a charge) arrange to have the supplier truck the goods to your home.
- When transporting or unloading sheet goods, try to have a helper on hand. If that's not possible, grab hold of the panel as shown in the drawing inset below (one hand near the center of each long edge) then pick it up and wrestle it to its destination. Take care that you don't damage the edges or scratch the surface of the panels. And keep in mind that drywall, because it's thin, heavy, and brittle, can snap under its own weight.
- Store all materials in a cool, dry place, off the floor. Moisture can distort lumber, delaminate some plywoods, and render drywall useless.
- If possible (often it isn't), store sheet goods flat. If lack of space prevents this, stand them on one of their long edges, as nearly vertical as you can.
- Store lumber flat, and weight it down at each end and in the center to prevent warping and other distortions. Weighting down is especially important if you purchased wood with high moisture content.
- To keep excess building materials from cluttering up your shop, you may want to build a storage rack (the one shown below is easy to construct and occupies a minimum of space).

Selecting Hardware

Nails

If you've always thought of nails as plain Jane fasteners, the sketch below may surprise you. It shows a grouping of commonly sold types—each one carefully engineered for a specific use.

Both *common nails* and the thinner-shank *box nail* excel at construction-type work (use commons for heavy jobs, box nails for lighter work). Use *roofing nails,* which have wide, thin heads, to hold down composition roofing.

Casing nails, finishing nails, and *brads* can tackle heavy-, medium-, and light-duty finish work, respectively. For a neat appearance, you should countersink them, then fill the depressions.

Ring-shank and *spiral nails* are designed to achieve a firm grip in wood. Specially hardened *masonry nails* can penetrate mortar joints and concrete. The protruding top head of *double-headed nails* allows quick dismantling of temporary work. *Corrugated fasteners* are used mainly for strengthening wood joints.

To determine the size nail you need, keep in mind that generally you want one long enough to go two-thirds through the material you're nailing into. Once you've determined the correct length, refer to the size chart below. Many, though not all, nails still are sold by the antiquated "penny" system, so if you need 3½-inch common nails, for example, ask for so many 16-penny common nails. You can buy nails either in bulk or already boxed.

For most projects, you'll be using uncoated steel nails, though they are available in several other metals as well. For outdoor use, buy galvanized steel nails. One other coating deserves mention, too. So-called *cement-coated nails* have an adhesive coating that actually bonds to wood fibers under the heat and friction of driving. Use these when superior holding power is a must.

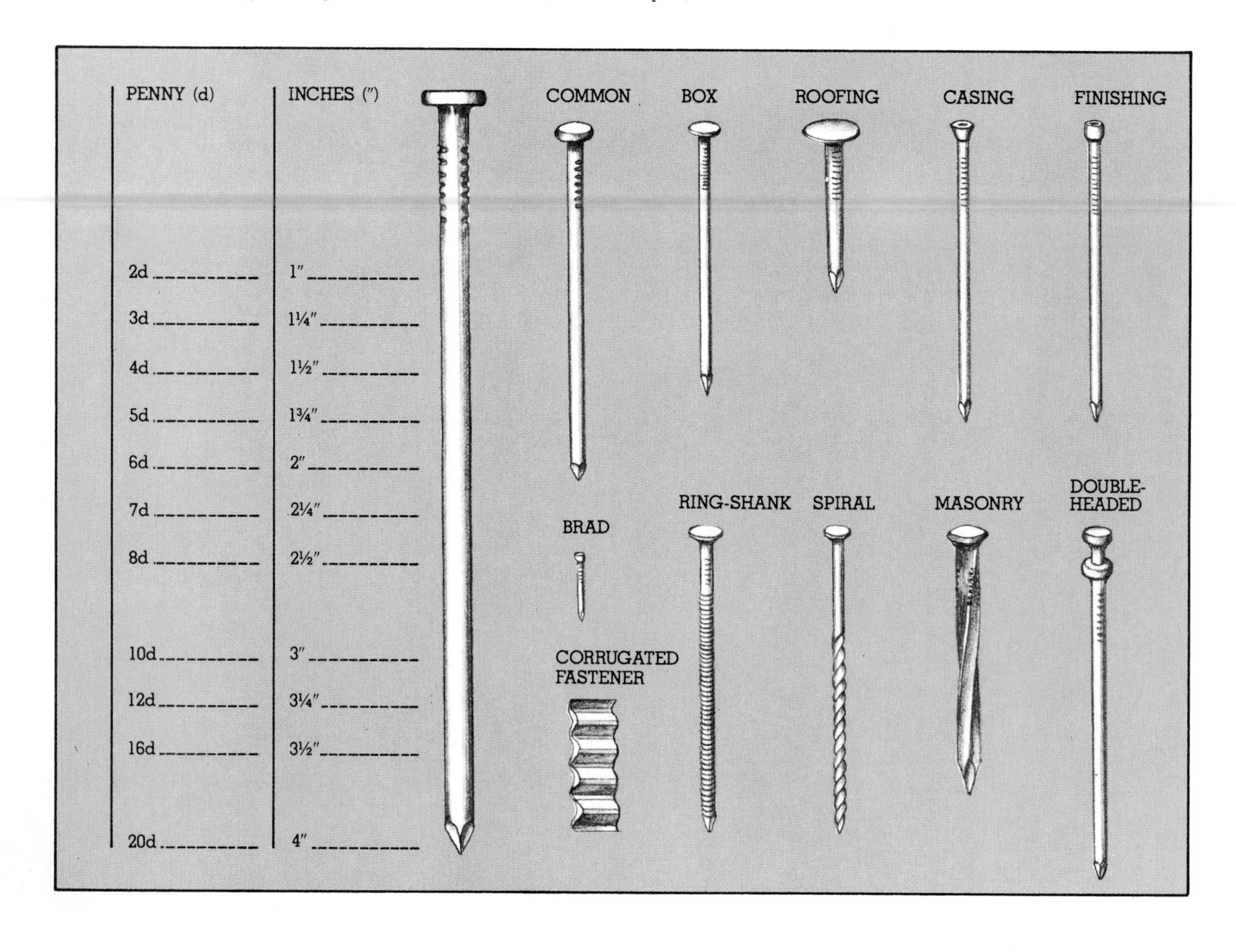

Screws and Bolts

For the few seconds they take to drive, nails do a remarkable holding job. Yet for the extra time involved in driving a screw into the same material, you get an even tighter hold, a neater appearance, and another plus—the option of disassembly.

The sketch below shows, among other items, the two broad types of screws you'll run across—*wood screws* and *metal screws.* Wood screws, the type you'll be using most often for carpentry work, come with various head shapes *(flathead, ovalhead,* and *roundhead)* and slot configurations *(single slot* and *Phillips).* And they are available in plated steel and brass. Use flatheads when the screw must be flush with the surface, ovalheads for decorative accent, and roundheads for more utilitarian tasks. *Flat* and *trim washers* protect against marring wood surfaces.

As with nails, the right size screw for the job should go two-thirds through the member you're fastening to. When ordering wood screws, specify the *length* (from ¼ to 5 inches), *gauge* or shank diameter (No. 0, which is about 1⁄16 inch, to No. 24, about ⅜ inch), *head type,* and material. The larger a screw's gauge (that is, the thicker its shank), the greater its holding power.

Use *lag screws* for heavy framing jobs and *hanger screws* for hanging heavy objects. To order lag screws, specify *diameter* (from ¼ to 1 inch) and *length* (from 1 to 16 inches).

Machine and *carriage bolts,* both of which usually are zinc-plated to resist corrosion, handle all kinds of heavy fastening jobs. Which type you use isn't critical.

When ordering bolts, specify the *diameter, length,* and the *type* (machine or carriage). Add about ¾ inch to the combined thicknesses of the materials to be joined to determine the correct length. The diameter you need depends on the strength required of the fastener. Carriage bolts are available in diameters ranging from ¼ to ¾ inch and lengths from ¾ to 20 inches. With machine bolts, choose from diameters of ¼ to 1 inch and lengths of ½ to 30 inches.

For information about two specialized hollow-wall bolts—toggle bolts and hollow-wall anchors—see pages 56–57.

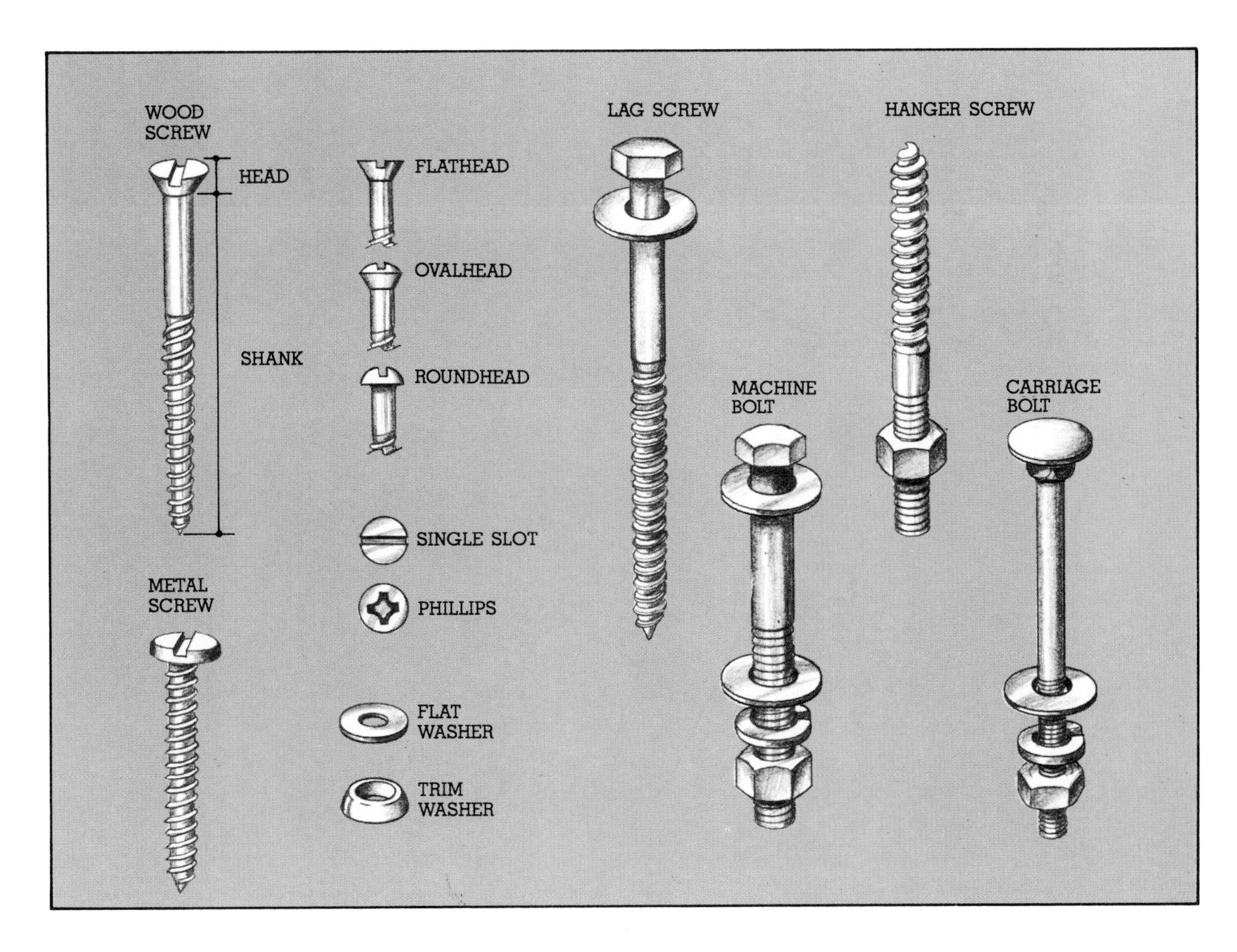

Miscellaneous Hardware

The grouping shown on these pages represents just a few of your specialized hardware options. The best way to get a feel for the multitude of items available is to spend some time in the hardware section of any building materials outlet. There you'll find a product designed for every conceivable carpentry need.

Shelf Supports

Aluminum shelf standards and brackets, your most attractive and versatile choices, come in a wide range of sizes and finishes. A lip on the front edge of each bracket holds shelves in place.

Less attractive but quite functional, bent-metal *Z brackets* and *utility shelf brackets* support shelves in a fixed arrangement. With utility brackets, fasten the longer leg to the wall. Fasten both types to walls with nails or screws.

When closed, *folding steel brackets* hug the wall. But when locked in their open position, they provide firm support for shelves or work surfaces that need periodic setup and dismantling.

Metal Plates and Framing Fasteners

Whenever you want to beef up an otherwise weak wood joint, reach for one of the four plates shown below. And for heavy framing jobs—stud walls, joists, rafters, and so forth—you may want to consider using framing fasteners, one of which (a *joist hanger*) is shown. *Mending plates* reinforce end-to-end butt joints; *T plates* handle end-to-edge joints. *Flat corner irons* strengthen corner joints by attaching to the face of the material. *Corner braces* do the same thing, but attach to edge surfaces.

Door and Cabinet Hardware

Most full-size doors, and some smaller ones, too, hang on the classic *butt hinge. Continuous hinges* flush-mount on cabinets and chests, combining great strength with a slim, finished look. *Strap* and *T hinges* often appear on gates and trunk lids.

You'll run into three basic types of cabinet hinges, too. You can use *offset hinges* on lipped, overlay, or flush doors, and *pivot hinges* for doors that completely overlap the frame. For cabinet doors set flush with the frame, choose a surface-mounted *decorative hinge* or the already-mentioned butt hinge.

To open your cabinet doors; fit them with *knobs* or *pulls,* available in myriad sizes and styles. *Roller* or *magnetic catches* keep them closed (if you use self-closing hinges, these aren't necessary).

And to keep drawers on track, you'll need *drawer slides,* either side- or bottom-mounted. The ones shown opposite use nylon rollers and ball bearings in side-mounted tracks.

Adhesives

Although not "hardware" as such, adhesives do play an important role in many fastening jobs. Lay in a supply of *wood glue* for general-purpose interior work and *epoxy* or *waterproof resin glue* for exterior jobs. You may also want several tubes of *panel adhesive* if you'll be installing drywall or paneling.

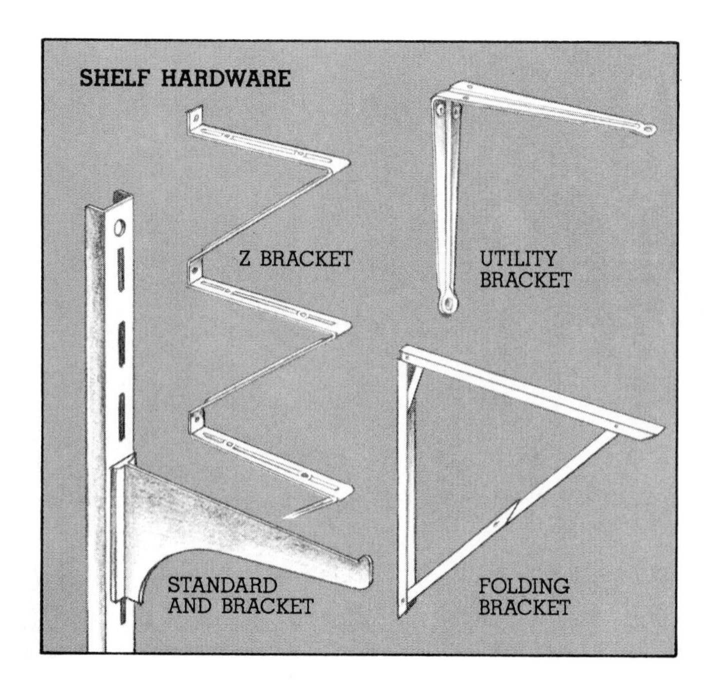

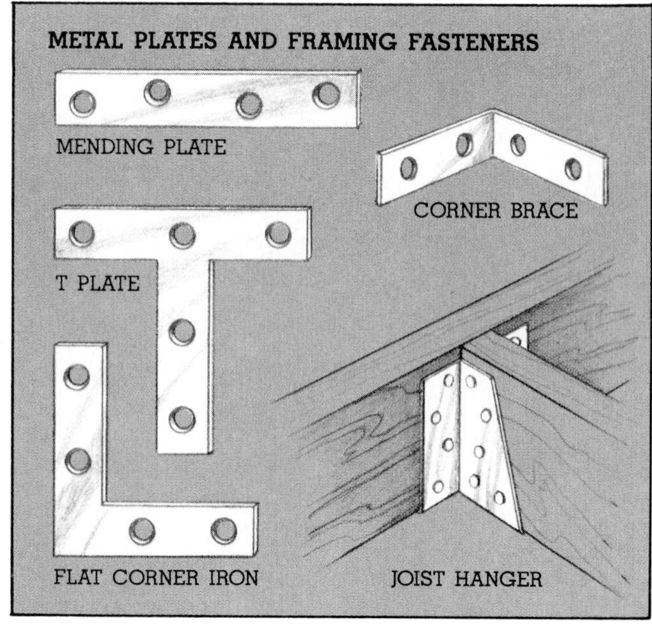

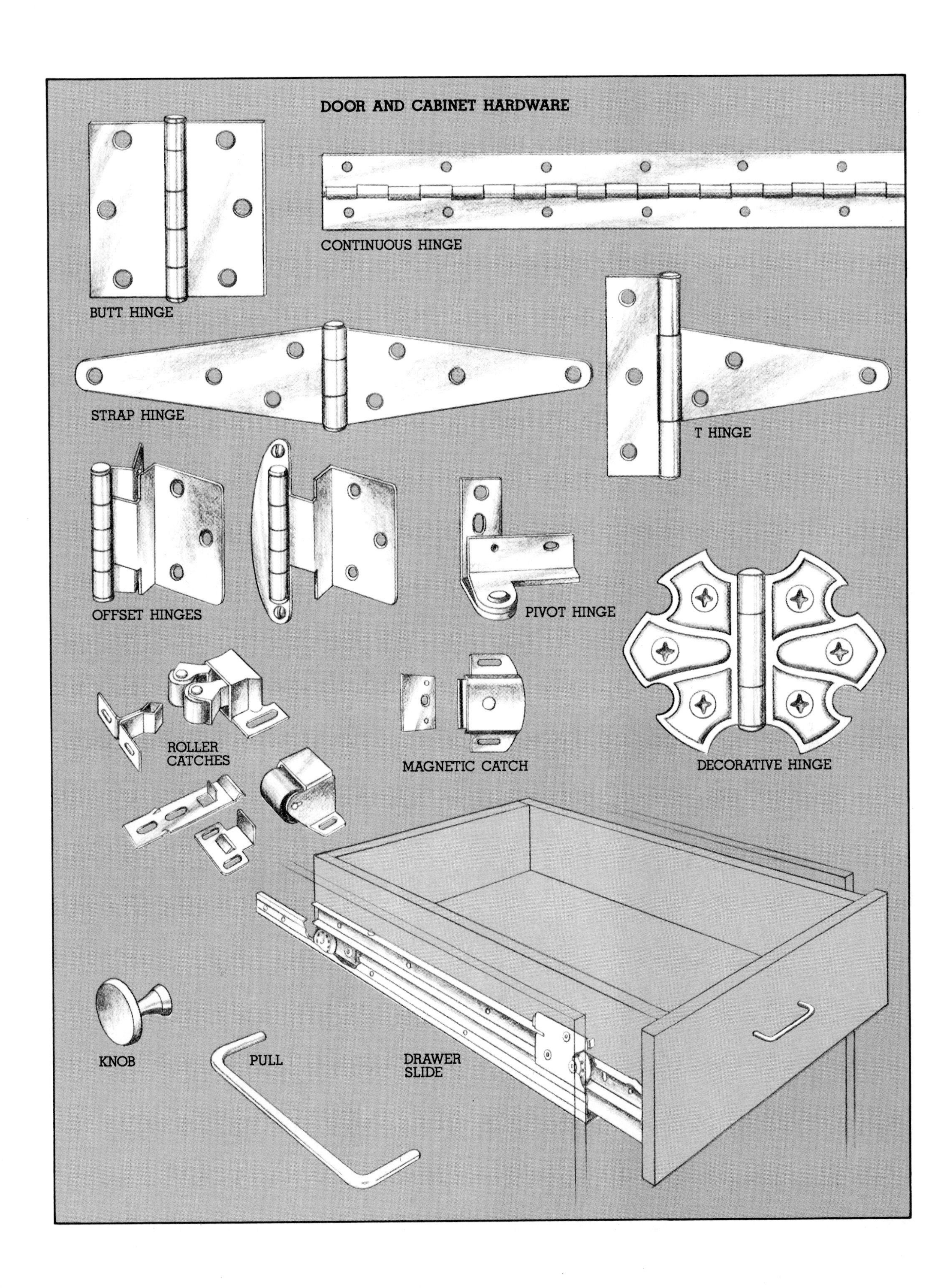
DOOR AND CABINET HARDWARE
BUTT HINGE
CONTINUOUS HINGE
STRAP HINGE
T HINGE
OFFSET HINGES
PIVOT HINGE
ROLLER CATCHES
MAGNETIC CATCH
DECORATIVE HINGE
KNOB
PULL
DRAWER SLIDE

MASTER THESE BASIC TECHNIQUES

If you can't quite envision yourself as a capable home carpenter, try this. Think of each of the techniques presented in this chapter as a building block. And each project you tackle as nothing more than several of the blocks stacked up. That's all any carpentry project is—no matter how simple or elaborate.

But also realize that just knowing how to do something is quite different from doing it well. What we provide here are the techniques. The rest is up to you, and that means practice, practice, and more practice.

Note that we've arranged the techniques so the order coincides with the sequence in which you'd do them when working on a project.

We open with a survey of measuring techniques, the mastery of which is imperative if you hope to achieve good results. Then it's on to cutting, drilling, fastening, shaping, smoothing, and finishing techniques.

And because it's nearly impossible to do satisfactory work with dull tools, there's a brief section on tool sharpening.

Measuring Techniques

Square the Work

Have you ever very painstakingly marked a material for cutting, carefully followed your cutoff line, and still wound up with a piece that didn't fit correctly? You probably made the mistake of beginning your measurement from an edge that wasn't *square*—at a 90-degree angle to an adjoining edge. Follow our tips for starting out square (and staying that way), and you won't make that mistake again.

1 Check board ends for square by positioning a combination square or try square with the body or handle firmly against a factory edge. (For more accuracy use a framing square.) If the end isn't square, mark a line along the outside edge of the square's blade.

2 To determine whether adjoining members are square, lay a framing square up against both members where the two meet. If the tongue and blade of the square rest neatly against the members, all is well.

3 No framing square handy? Then just call back some of your high school geometry and lay out the two sides of a 3-4-5 triangle at the test corner, using whatever units are convenient. If the corner is square, the diagonal joining the sides will measure five units.

See the detail for a third equally satisfactory way to check a project for square. This technique involves measuring diagonally between opposite corners. If your layout is square, both measurements will be the same.

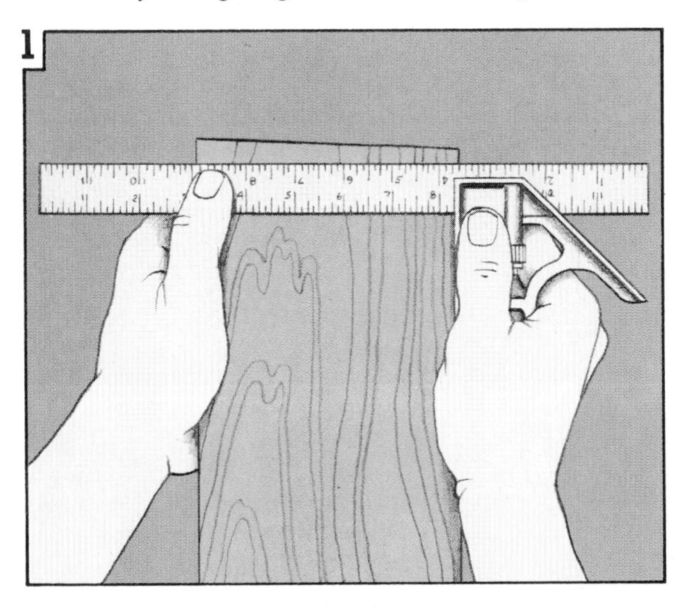

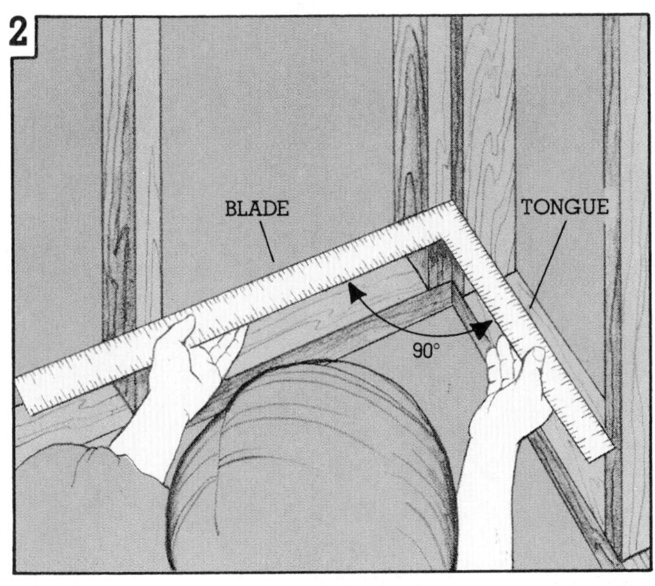

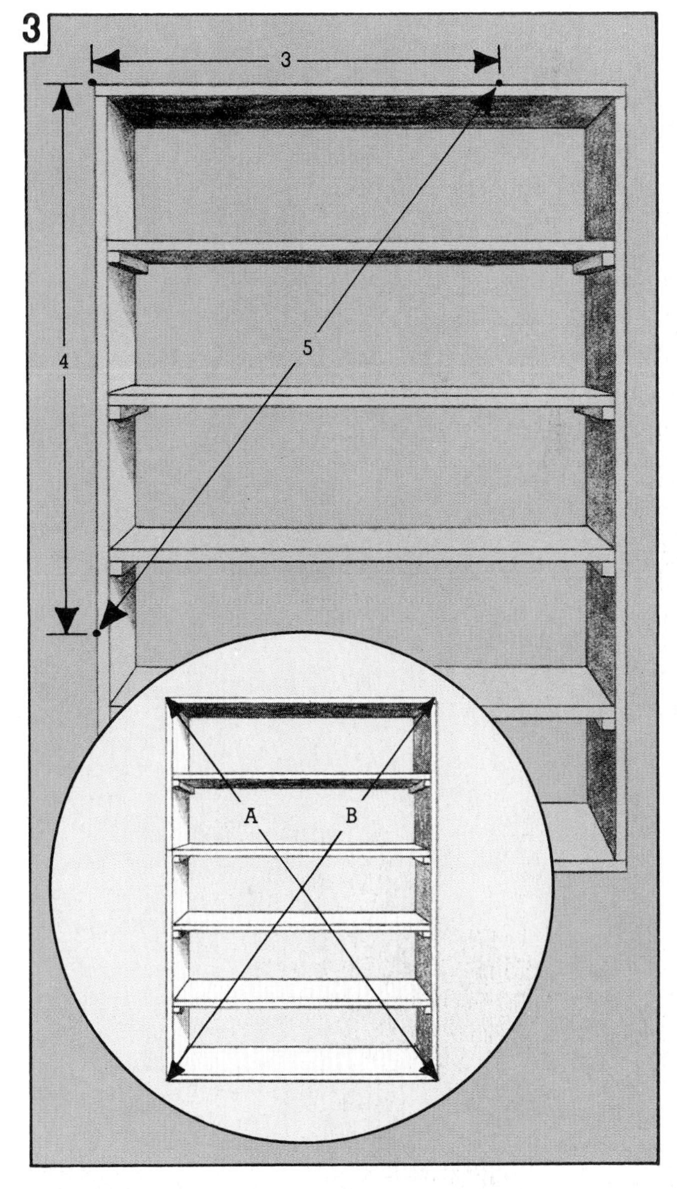

Mark with Care

How important are accurate measurements? VERY! Just ask a building contractor who has just miscut a rafter, or a cabinetmaker who suddenly realizes that a drawer opening is a glaring ⅛ inch too wide. To these pros, and to you, too, being off on a measurement means lost time and money—not to mention plenty of irritation.

When an experienced carpenter makes a measuring mistake, it's generally because he's in a hurry. The same holds true of a beginning woodworker. So, especially when you're just starting out, *take your time* and adopt the policy, "Measure twice, then cut once."

Also, no matter what measuring device you use to position your marks, learn to read it accurately. Many a board has met its ruin because someone couldn't distinguish a ¼ inch from a ⅛ inch.

And once you have made a measurement, jot it down on a piece of paper or a wood scrap. This way you can refer to it later, eliminating the need to remeasure.

Last but not least, use a sharp instrument for marking. A sharp No. 2 pencil works well, as does a scratch awl or a knife. Shy away from flat carpenter's pencils; their bulky size can cause you to make mistakes.

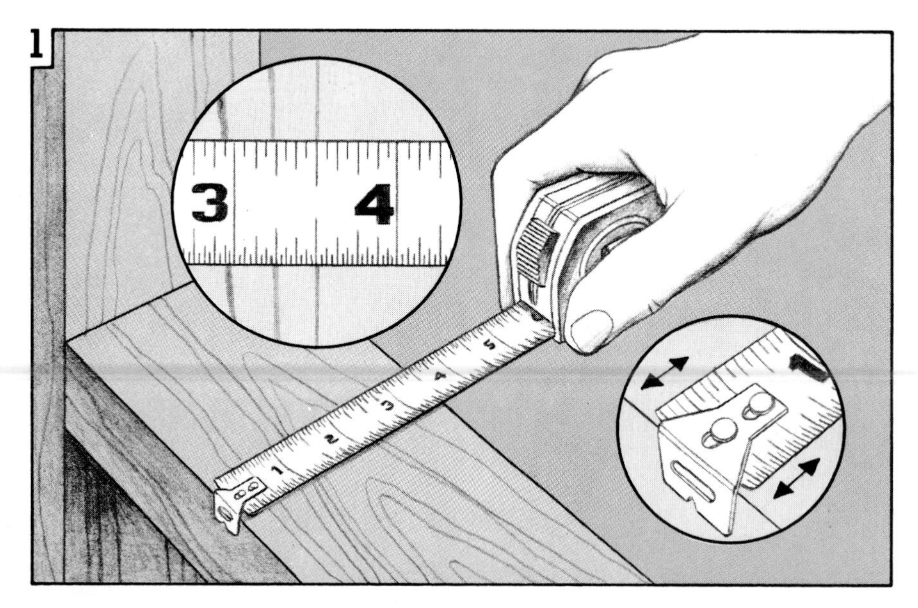

1 By far the most versatile measuring device, the steel measuring tape does lots of measuring and marking jobs well. Here, it's taking an outside measurement. Note how the hook at the end of the tape slides out to compensate for its own thickness. Note, too, that along the bottom of the first several inches, each inch is divided into thirty-seconds to facilitate extra-fine measurements.

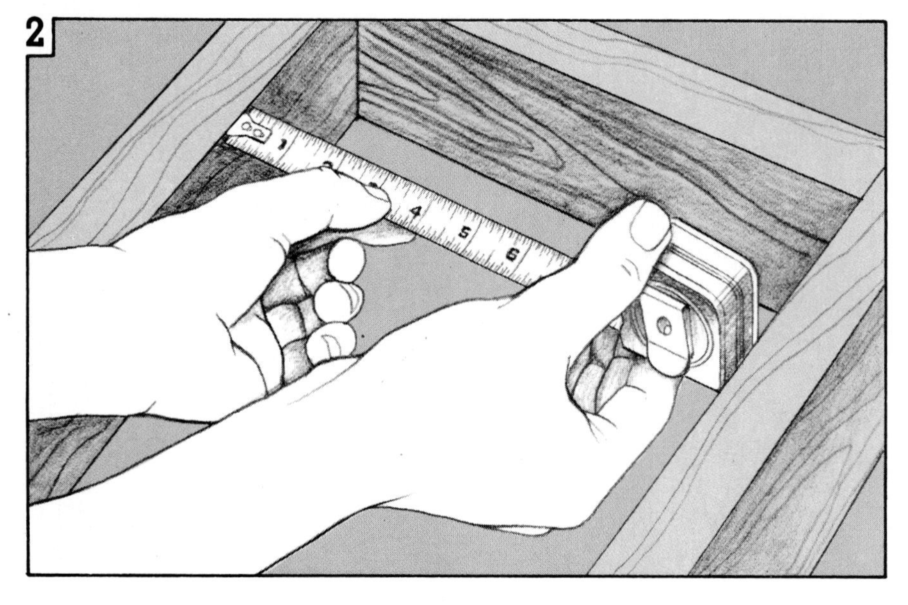

2 Here's the same tape, this time taking an inside measurement. Simply add the length of the tape case (it's marked on the side of the tool) to the length of the extended blade and you have the total inside dimension. Some newer models do the math for you and give readings through a small window on the case top.

3 Marking, as you can imagine, involves just as much care as does measuring. Mark all your cutoffs with a V so you know precisely where to draw the line. And to ensure pinpoint accuracy, place the point of your pencil at the V, slide the square up to it, then strike your line. (To extend 45- or 90-

degree angle lines across most board widths, you'll find a combination square the ideal tool. For longer line segments, use a framing square or a straightedge.)

4 Need to mark a cutoff line along the length of a board or a piece of plywood? For a fast but slightly rough line, use your ever-handy measuring tape. Lock the blade at the dimension you want, angle a pencil into the notched clip, and pull it firmly toward you as shown.

For sheet goods, first mark the cutoff line at both ends of the sheet, then snap a chalk line between the two marks. Or make your marks, then use a straightedge to complete the line.

5 Whenever you cut any material, the saw blade reduces some of it to sawdust. Because of this, you must allow for the narrow opening in the blade's wake—called the *kerf*—when making your measurements. Otherwise, you could end up short 1/16 inch or more, depending on the kerf's width. If you're making just one cut, account for the kerf by identifying the scrap side of the cutoff line (by X-ing it as shown) so there's no confusion later about which side of the line to cut from. For multiple cuts along a board or on sheet goods, allow for the width of the saw kerf between the kerf lines.

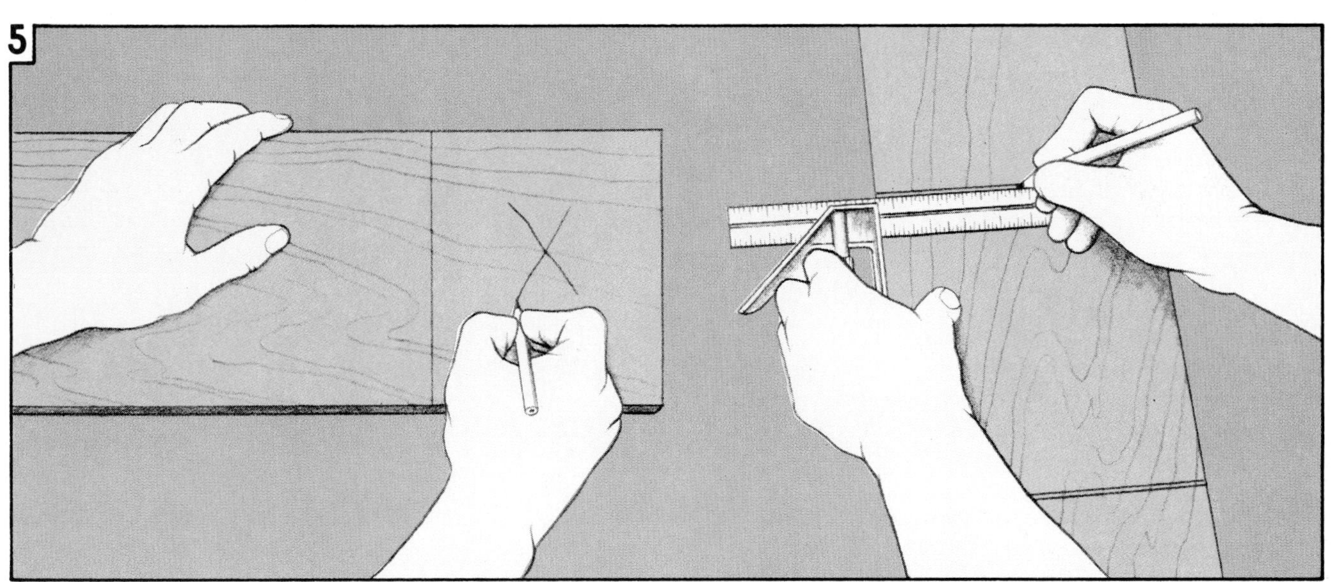

Keep Projects Plumb and Level

Ask any three carpenters how to drive a nail correctly, the best way to cut roof rafters, install a pre-hung door, or almost any other carpentry-related question, and you'll probably get three different answers. That's because in carpentry, there often are several "right" ways to do something. But one thing these pros all agree on: the importance of making sure that projects are plumb and level, as well as square, which we discuss on page 27.

In case you're not familiar with what plumb and level mean, a surface is plumb when it is vertical, and level when it is horizontal.

To understand the importance of plumb and level, just consider what happens if you don't bother checking for them. The results are predictably unsatisfactory—walls lean, shelves slope or tilt, newly installed doors bind in their frames, to name just a few maladies.

As you're about to discover, checking for plumb and level is neither difficult nor time consuming. So don't sacrifice the appearance and function of a project for the few seconds it takes to make these vital checks.

1 To check most projects for plumb, hold a carpenter's level against one face of the vertical surface—a partition wall in this instance—and note the position of the air bubble in the appropriate glass vial. If it comes to rest between the two guide marks printed on the vial, you know the wall, for instance, is plumb. If it doesn't, tap the project the direction needed to achieve plumb.

Vertical members must be plumb in two directions, so you'll need to repeat this procedure on a surface adjacent to the one you just checked.

2 On occasion you may need to locate a point on the floor directly below a point on the ceiling, as when determining the placement of the top plate and soleplate for a wall, for example. A level would not help here, as you have no surface to support it on. Instead, use a plumb bob—a pointed weight suspended on a line.

Simply attach a line to the overhead reference point so the bob falls just short of the floor. When the bob comes to rest, mark its position with intersecting lines. In the case of a wall, you'd repeat the procedure several times.

3 You can check most projects for level by simply setting your carpenter's level atop the member—here, we're leveling a closet shelf—and raising or lowering the member until the level's bubble rests between the register marks. Then mark the member's position for later reference.

4 A member must be level from *front to back* as well as from *side to side,* so be sure that you make this check, too.

In quarters too tight for a carpenter's level, use a torpedo level if you have one or the level on your combination square.

5 Sometimes, you'll need to establish level over a long distance where there's no convenient support for a carpenter's level, as when cutting fence posts or deck supports to size. Here, go with the versatile line level.

To set up a line level, fasten a heavy string to your reference post and pull it taut to the post you want to level. With the line level hung midway between the posts, mark the post for cutting at the height where the line is level.

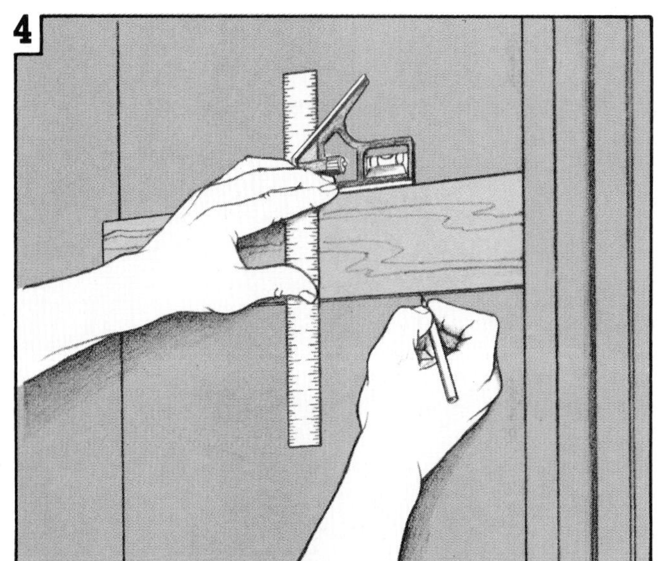

Special Instruments for Special Jobs

If carpentry involved nothing more than straight line after straight line, a flexible steel tape, a square, and a level would suffice for most measuring needs. But carpentry isn't always quite that simple. Angles, curves, and circles all present special situations best dealt with by using the tools shown on this and the following page.

As with all other carpentry techniques, the more accurate your measurements are, the better the finished product will be. So let's take a look at how each of these will help you get the job done right—the first time.

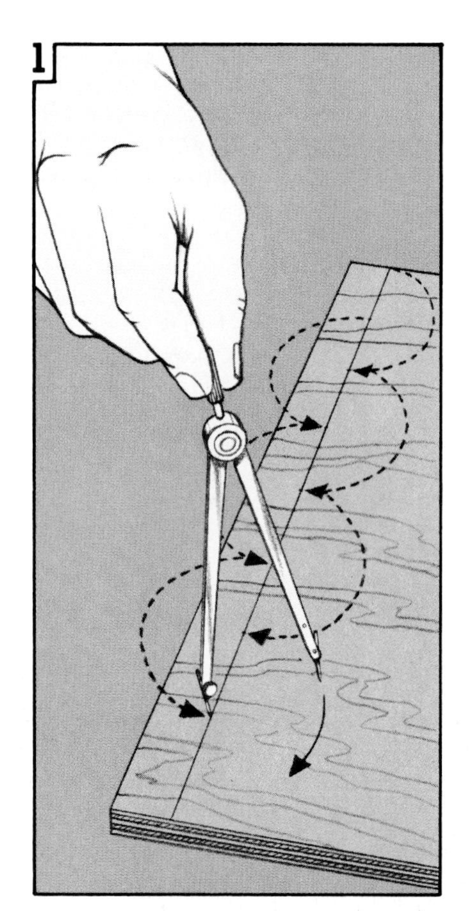

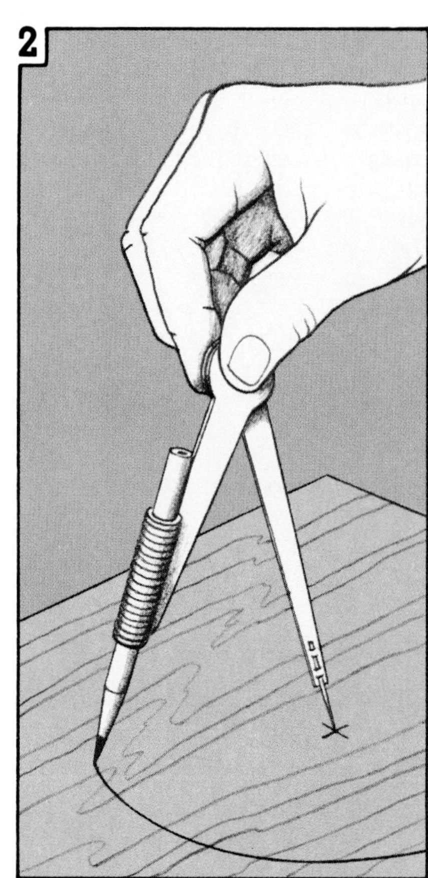

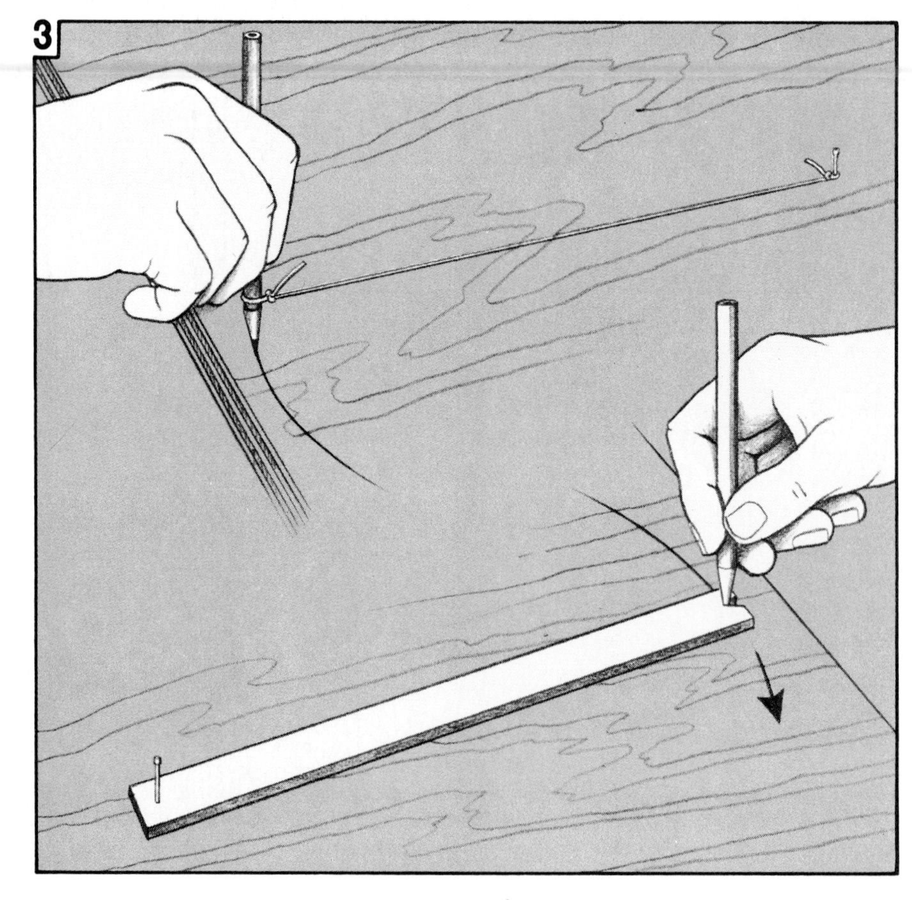

1 For stepping off equal distances along a straight line, it's hard to match the pinpoint accuracy of *wing dividers*. Steel pins at the base of each leg grip the surface of the material you're measuring for positive control. Note the rocking motion used here to walk the dividers across the plywood.

2 Widely acknowledged as one of the workhorses among measuring/marking tools, a *compass* will perform an array of tasks well. Here it's outlining a circle, but you also can call on it when you need to transfer contours, step off distances, or measure short inside or outside distances. When making circles with a compass, hold it almost perpendicular to the surface, as shown in the sketch.

3 As mentioned above, for making small circles a compass is the tool to use. But what about those large cutouts such as lavatory openings in counter tops? In this and similar situations, fashion your own compass, using a pencil, heavy thread or a thin piece of wood, and a brad or other small nail. Nail the brad at the center of the circle and, with the pencil perpendicular to the surface, scribe your circle as shown.

Too slow for you? Then find a can, coin, ashtray—anything fairly true and close to the size circle you want—and trace around it.

4 *Calipers,* though admittedly measuring-tool extras, come in handy for many tough jobs. To use *inside calipers,* fit the tool within the opening you want to measure, tighten the setscrew, withdraw the tool, and then read the distance against a measuring tape. *Outside calipers* record various outer dimensions such as the diameters of pipes, conduit, and rods. Another tool, the *caliper rule* (see the detail) will measure both inside and outside distances.

5 Duplicating tricky angles can be one of the most hard-to-accomplish aspects of any carpentry project. But not if you have a *sliding T bevel* to guide the way.

When making an inside measurement with this tool, be sure to extend the blade fully to ensure a true reading. Once you're sure you have the correct angle, lock the blade in position by tightening the wing nut.

6 If you've ever tried to cut one material to match the contour of another surface, you'll appreciate what a *contour gauge* can do. It's simple to use. Just fit its movable metal teeth to the desired surface, then use the resulting shape as a template.

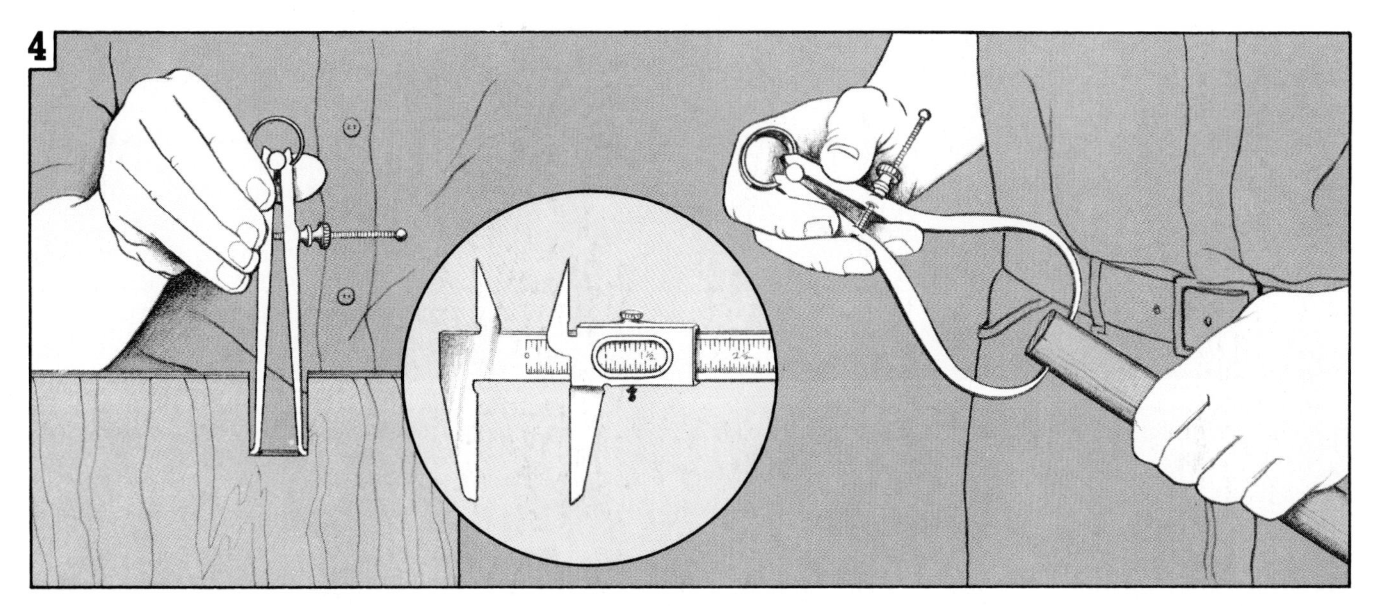

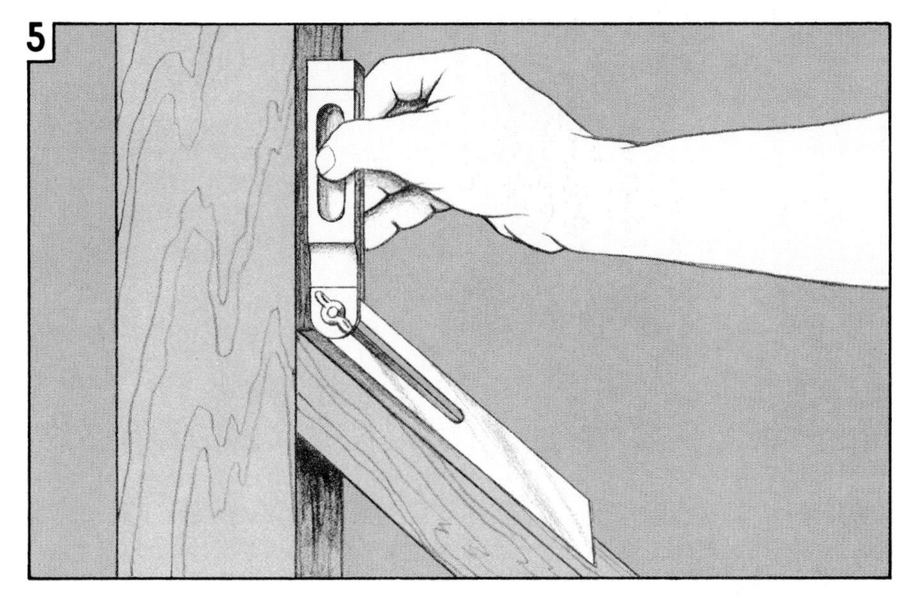

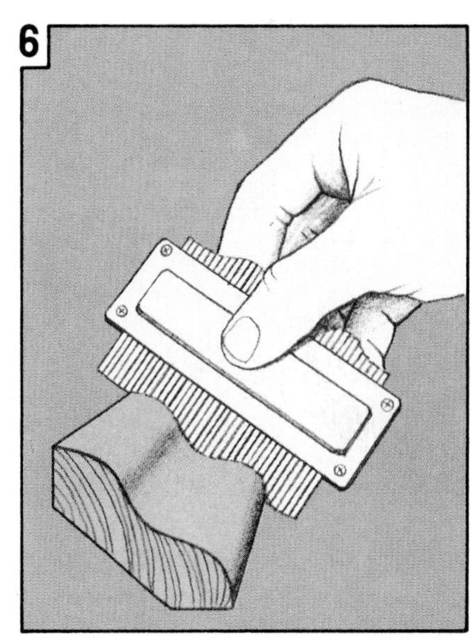

Cutting Techniques

1

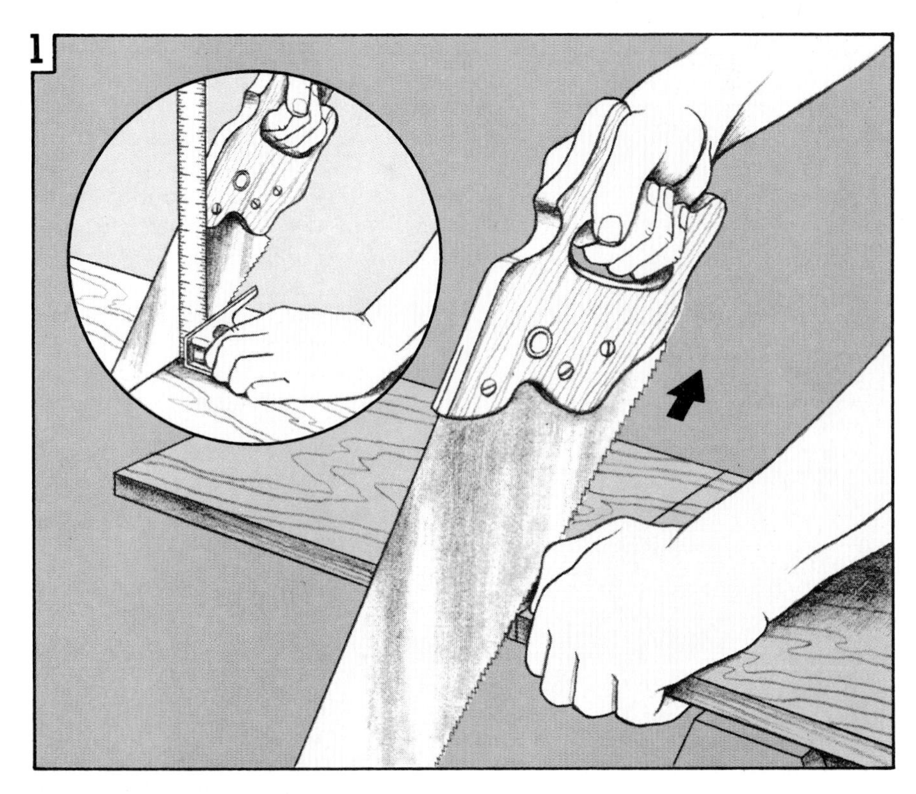

2

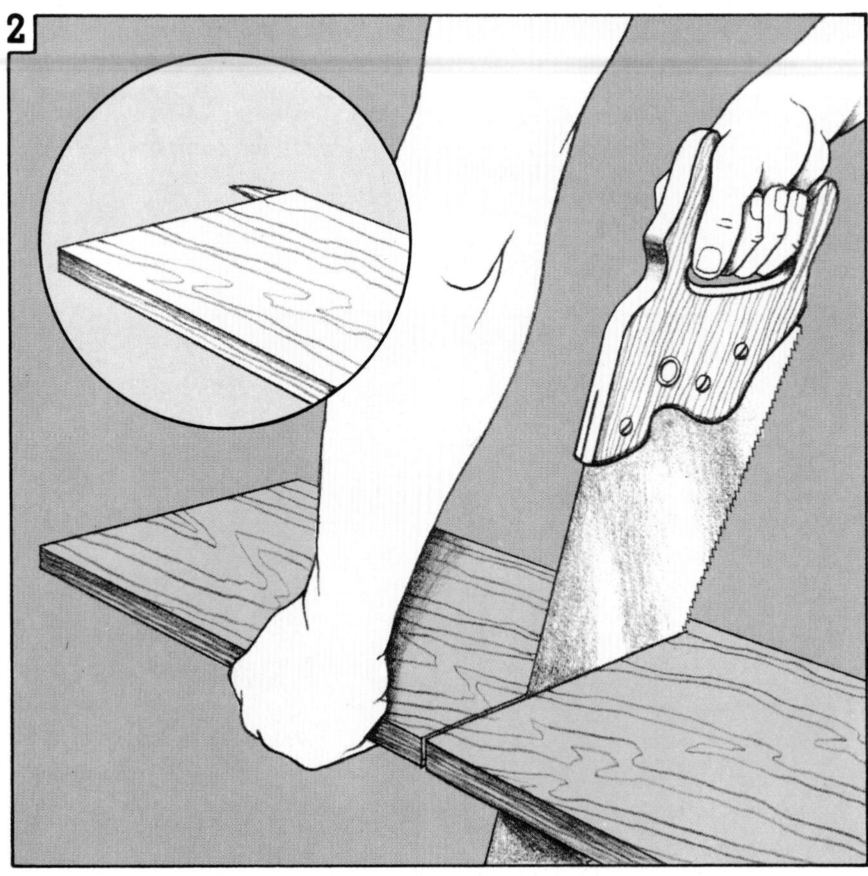

Making Rips and Crosscuts

Though from time to time you'll need to make the specialty cuts shown on pages 36–39, the vast majority of your cutting chores will involve either crosscuts (cuts across the grain of a material) or rips (cuts parallel to the grain).

On these two pages we've included some pointers that will enable you to make both types of cuts in a variety of materials with both hand- and power saws—and be proud of the results when you've finished. Many of the techniques you'll learn about here apply also to the more advanced cuts we'll show you on the following pages.

(Note: When cutting sheet goods that have a "good side," as paneling and some other panels do, you must guard against splintering them. To do this, cut with the material good face up if you're using a handsaw, good face down for saber and circular saws.)

1 To make a crosscut in narrow goods with a handsaw, start by setting the blade's heel end (nearest the handle) at a 45-degree angle to the work on the scrap side of the cutoff line. To make sure the blade doesn't stray from this position as you begin your cut, use the knuckle of your thumb as a guide. Pull the saw back toward you several times to start the cut. Don't force the blade—just let the weight of the saw do the work while you guide it. Then begin sawing,using a "rocking" motion, with a steeper blade-to-material angle at the beginning of the downstroke, and a slightly less acute angle at its completion. Periodically, check your work with a square to ensure that the saw blade stays perpendicular to the board.

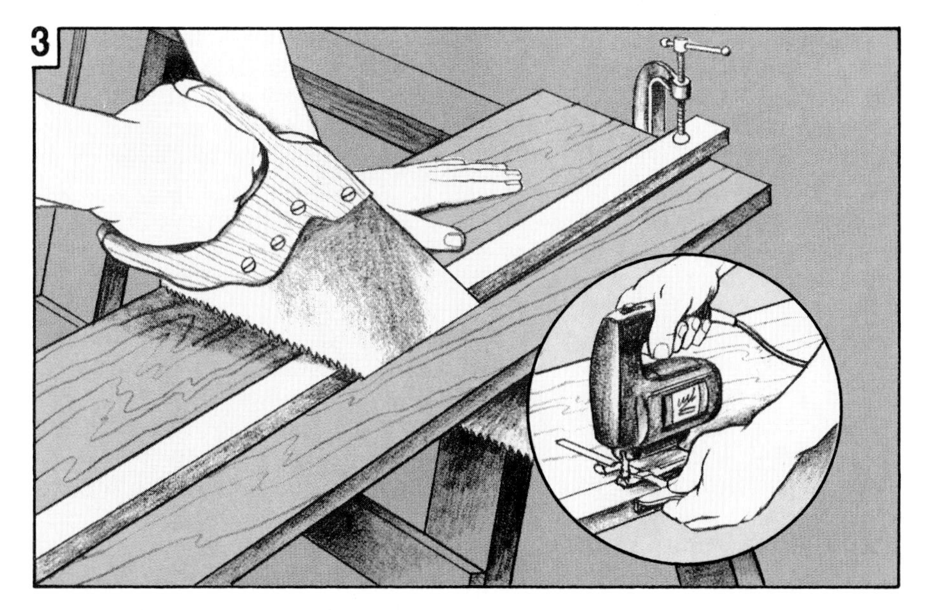

2 When you near the end of the cut, be sure to support the scrap piece of wood with slight upward pressure from your free hand. This keeps the piece from snapping (and splintering).

3 When ripping lumber or sheet goods, your main concern should be getting a straight cut along the entire length. To guarantee this, clamp a straight board or straightedge to the work to guide the saw. If the cutoff line is close to the edge of the material, you can accomplish the same thing with an edge-guide attachment fitted to a saber or circular saw.

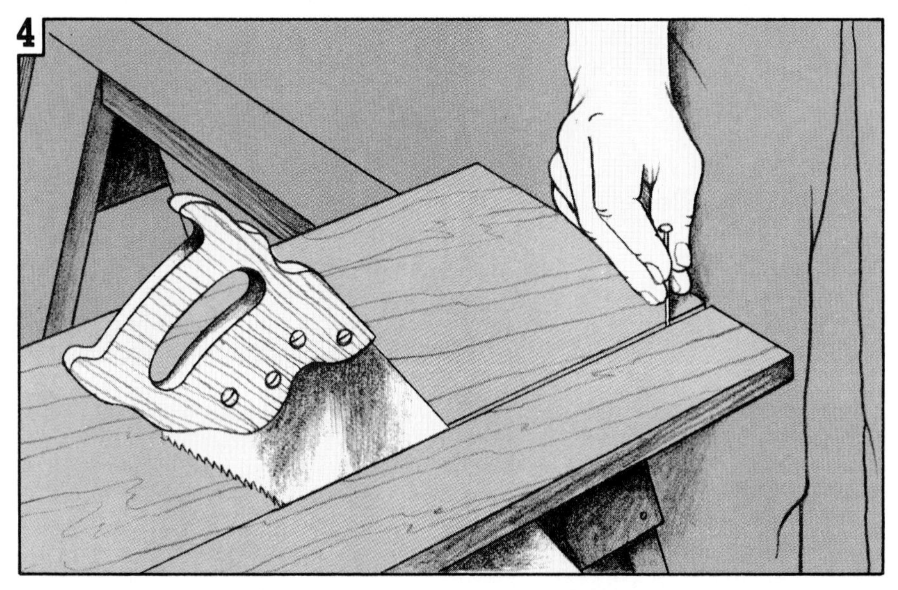

4 Because of the length of many rip cuts, saw blades tend to bind, especially near the end of the cut. If binding occurs, wedge a nail or a screwdriver into the kerf to keep it open, and give adequate support to the scrap piece.

5 Sheet goods are best cut with a circular saw that is fitted with a plywood-cutting blade. For this the trick is not so much one of cutting the material as it is supporting it properly. Here is one easy solution to this problem. Cut the sheet on a flat floor surface, supporting its entire length with 2×4s. Notice the two 2×4s on each side of the panel; they completely stabilize both halves during cutting.

Making Angle, Bevel, and Coped Cuts

From time to time, we've all marveled at the carpentry skill displayed in the joinery of a custom cabinet or hutch adorned with a variety of moldings—in a picture frame with its well-tailored, tight-fitting mitered corners, or in a seemingly seamless joint between two pieces of baseboard molding. But you know something? At the root of all this carpentry excellence is the ability to make accurate angle cuts, bevel cuts, and coped cuts—the three shown on these two pages.

Before getting into the techniques for making the cuts, let's briefly discuss what each is. When you make a cut *at an angle through the width* of a member, you've got an *angle cut.* A *bevel cut,* on the other hand, is one made *at an angle through the thickness* of a member. And *coped cuts* are those made to shape a member so it conforms to the shape of an adjoining member. (Note: When two members that have been angle cut or bevel cut at the same angle meet to form a corner, the joint is a *miter.*)

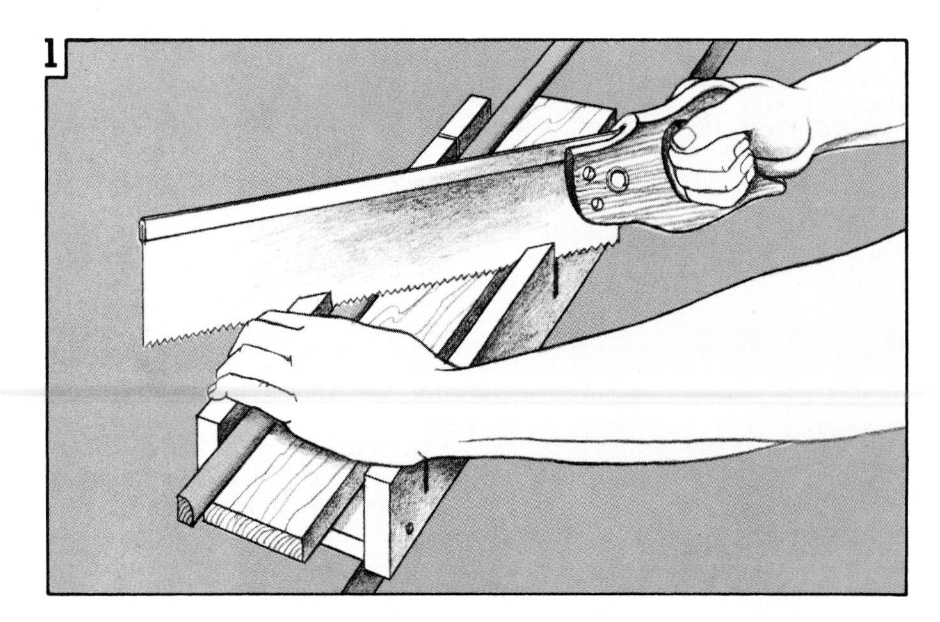

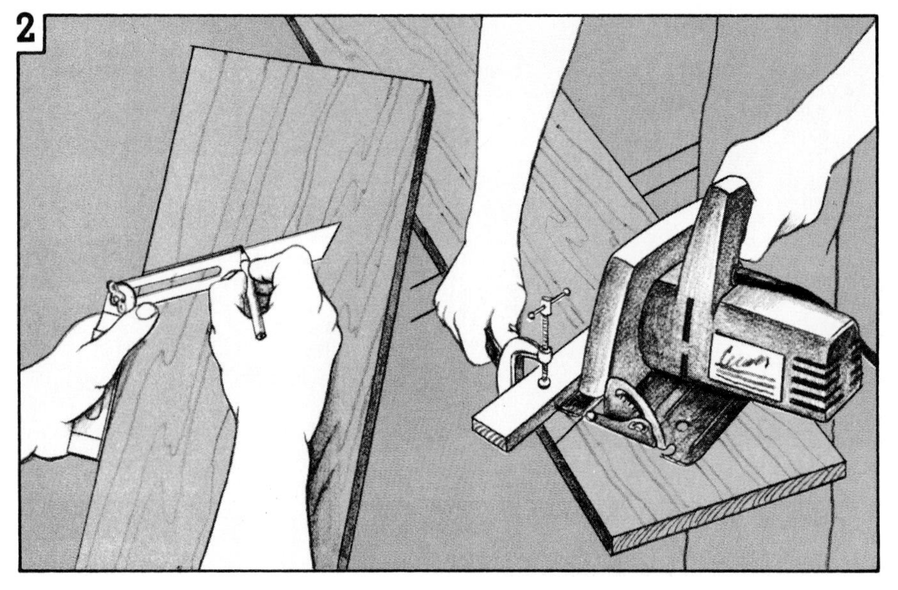

1 To make angle or bevel cuts in narrow stock the easy way, use a miter box—essentially a jig for holding the saw at the proper angle to the work.

Before placing the piece in the miter box, support it on a scrap of 1x4 or some other suitable material. This allows you to saw completely through the work without cutting into the bottom of the miter box. Position the member against the back of the miter box as it will be when in use, then use a backsaw to make the cut, holding the work against the back of the box with your free hand.

If there's any trick at all to using a miter box, it's not in the technique of cutting, but in correctly measuring and marking for the cut. To make sure you don't cut the piece too short or angle it the wrong way, draw a rough sketch of how the pieces will fit together before you begin.

2 As handy as a miter box is for making cuts through narrow stock, it just can't accommodate material much wider than a few inches. For angle cuts on material too wide to fit in a miter box, you can improvise a saw guide that will be just as accurate.

The first step is to set a T bevel to the desired angle and then to transfer that angle to the material. Now select a piece of 1x material with an edge you know is straight (or a straightedge) and C-clamp it along the cutting line as a saw guide.

To use a handsaw, clamp the guide close to the cutoff line. If you're using a circular or saber saw, offset the guide to position the saw blade alongside the cutoff line. To do this, measure the distance between the blade and the edge of the base plate, and clamp the saw guide this distance from the cutoff line.

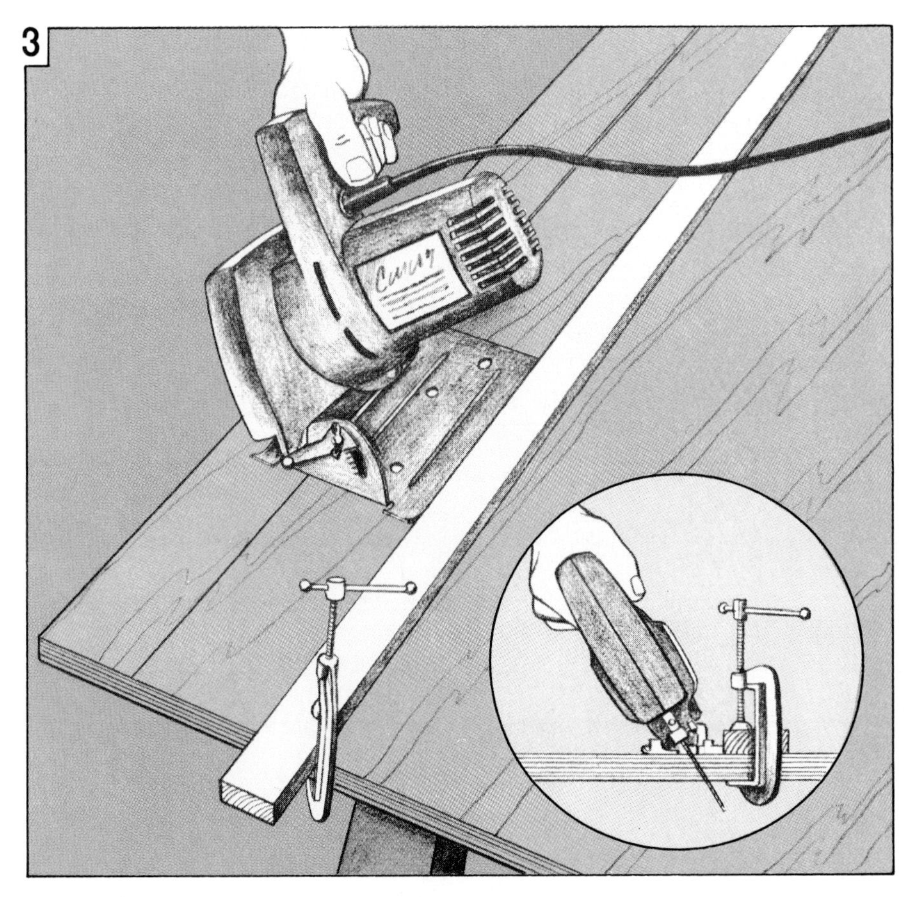

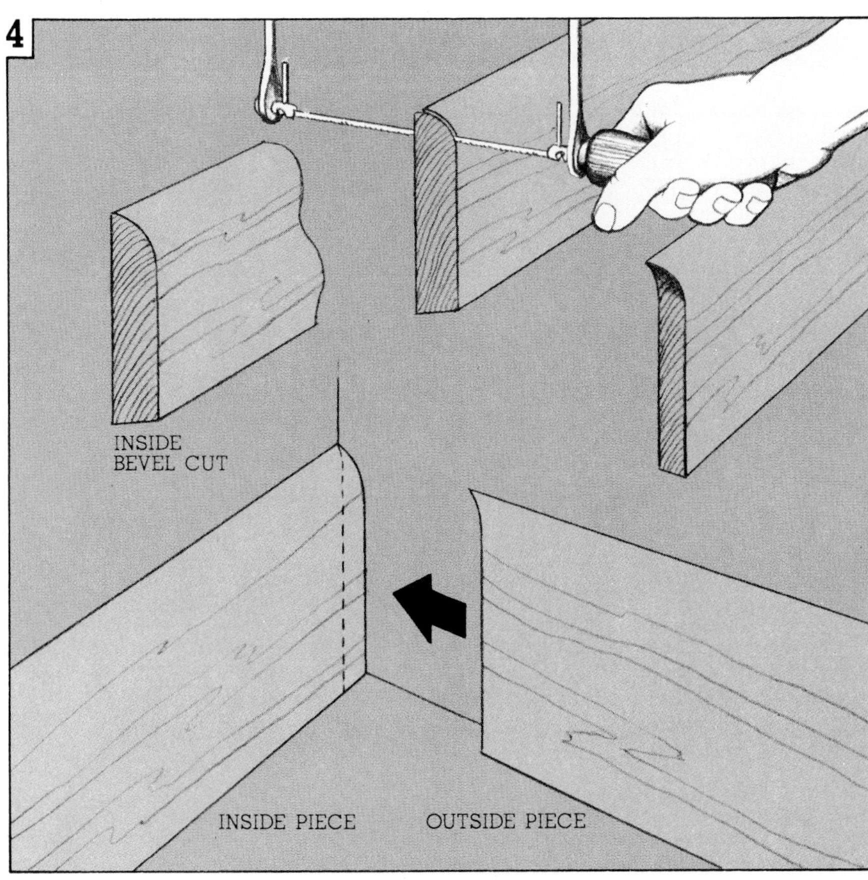

3 To make a bevel cut through wide stock, first use a straightedge to mark the cutting line on the face of the material. Then mark the exact angle of the bevel on the beginning edge. Now, with the saw unplugged (use either a saber saw or a circular saw), loosen the wing nut on the baseplate and adjust the plate-to-blade position to duplicate that angle. To double-check your work, place a correctly set T bevel between the bottom of the baseplate and the blade to verify the angle. Tighten the baseplate in this position, and make the cut by guiding the saw blade along the surface cutoff line.

4 When you're working with decorative moldings, it's sometimes hard to get a perfectly matched inside corner—especially if the corners aren't square to begin with. That's when it's useful to know how to cope an inside corner joint.

Start by cutting the "inside" piece of molding at a 90-degree angle so it butts against the adjacent wall. To cope the "outside" piece, first make an inside 45-degree bevel cut, as shown. Then use a coping saw to cut away the excess wood along molding's profile. Back cut slightly to avoid a gap if the walls aren't perfectly square.

Making Inside and Contour Cuts

Sometimes a carpentry project will throw you a curve—a cut you have to make from the center of your material, or an irregularly shaped cut. But don't be intimidated in the least. Neither inside nor contour cuts are difficult to make, though they do require some time and care.

1 The technique you use to start an inside cut depends on the material to be cut and the type of saw you're using. With lumber and most sheet goods, you have a couple of options. One way is to drill a starter hole at each corner of the piece to be cut away, as close as possible to the cutting lines. The other involves making a pocket cut with a saber or circular saw. To do this, first tip the saw forward on its baseplate as shown. Then start the saw and slowly lower the blade into the wood along the cutoff line. (You may want to practice this technique on a piece of scrap before making your finish cut. It takes some getting used to.)

Starting inside cuts in drywall is even easier. Just put the tip of a keyhole saw on one of the cutoff lines, and punch the saw's handle with the heel of your hand.

2 One advantage to inside cutting with a saber saw is that you *can* maneuver around corners easily and quickly. Don't try to turn the saw at a 90-degree angle, though. Instead, use the following two-step procedure for cutting corners perfectly square.

On your first approach to a corner, cut just up to the intersecting cutoff line. Now carefully back up the saw about 2 inches and cut a gentle curve through the scrap material over to that adjacent cutoff line. Continue in this manner, supporting the scrap material as you cut, until the scrap piece is free. Now you can easily finish trimming the corners with four short, straight cuts.

3 To make short work of most contour cuts, use a saber saw; its narrow blade lets you make curves and circles that are as smooth and perfect as you're able to draw them.

To make the decorative scallops on a window valance, for example, first clamp the board to a workbench or other firm surface, making sure nothing blocks the blade's path.

Before beginning the cut, allow the saw to reach full operating speed. Then as you make the cut, guide the saw slowly, without forcing the blade. If the saw begins to bog down or overheat, you're cutting too fast. Remember to support the scrap material as you reach the end of each cut to prevent it from breaking off.

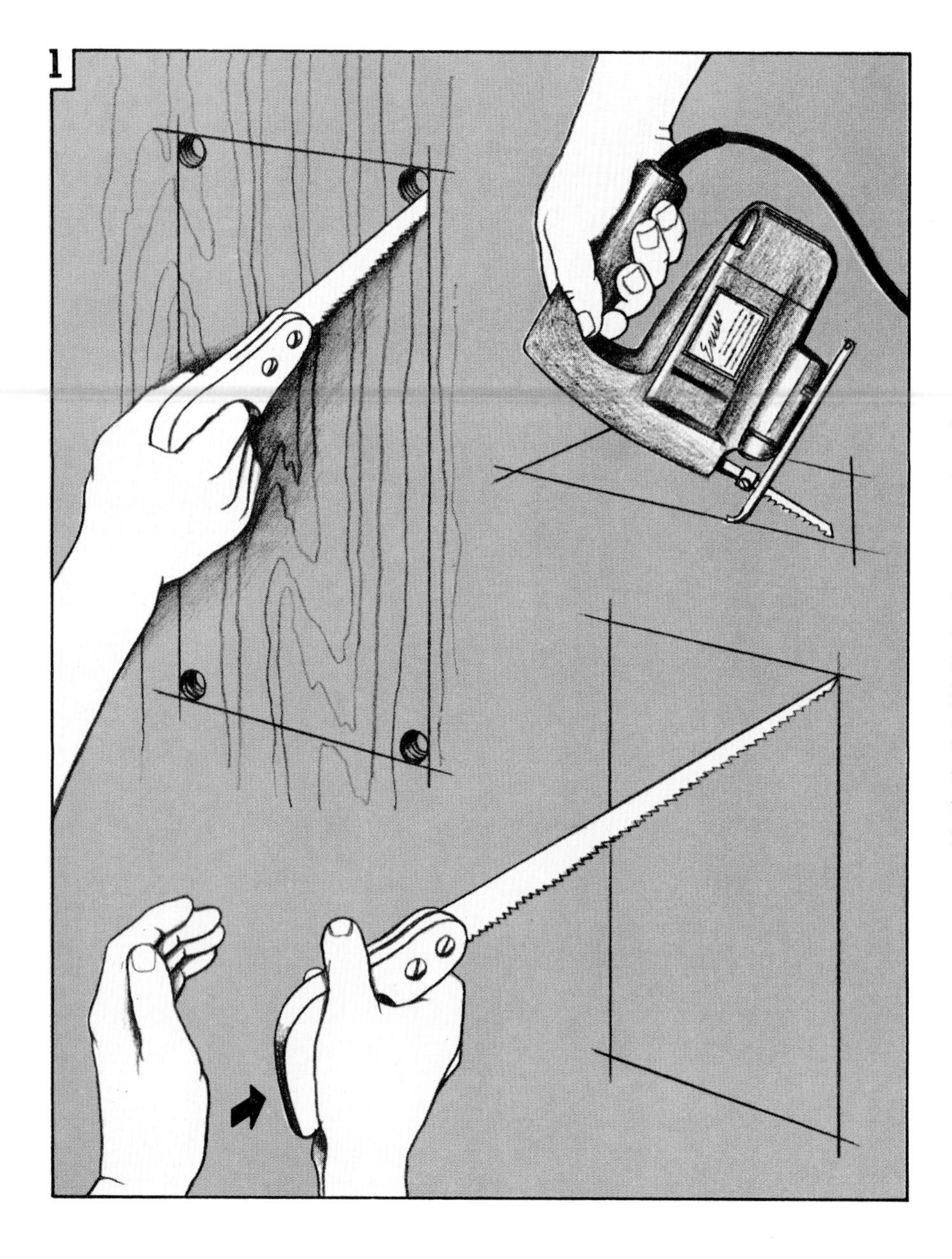

4 For intricate cutting or scrollwork jobs in narrow stock, you may want to use a coping saw. With one of these, you can approach the cut from an edge or from inside the material. To do the latter, just remove the blade from the saw frame and reinstall it through a starter hole.

For better control when making especially delicate cuts, fit the blade on the saw frame so the teeth are angled toward the handle. This enables the saw to cut on the backstroke. Blade holders conveniently let you turn the blade in different directions independent of the position of the frame.

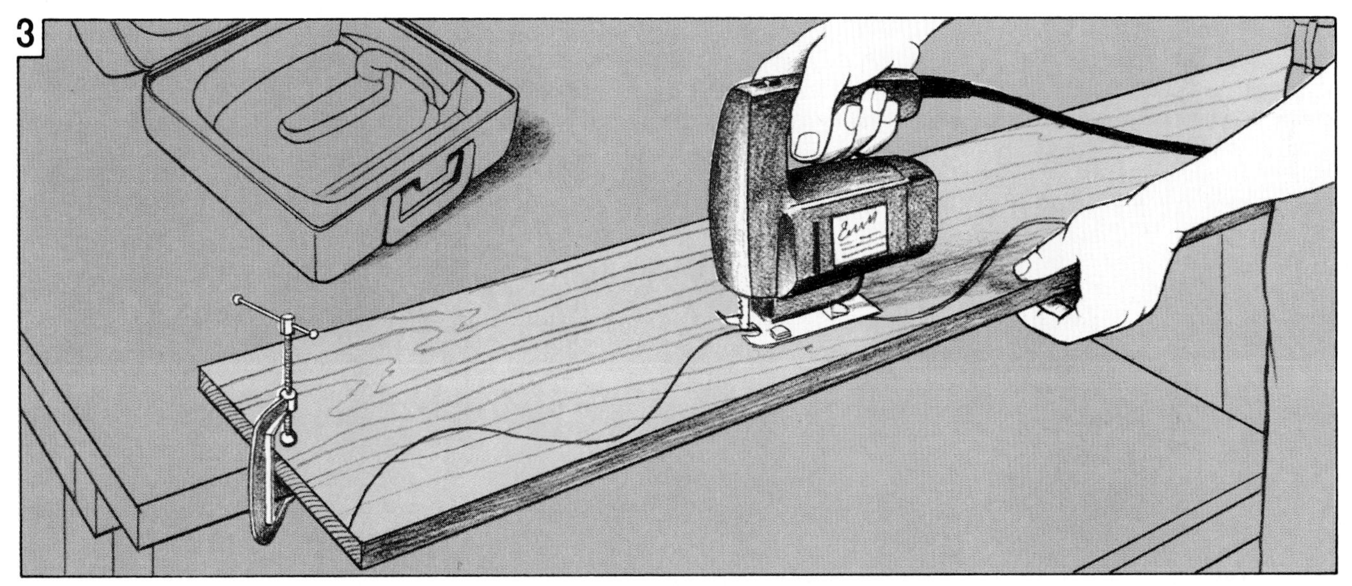

How To Use Chisels

Used to be, you could tell a true woodworking craftsman by his skill with a set of chisels. Few hand tools are as simple and, in experienced hands, as versatile. Today, routers do most of the work that once fell to chisels. Even so, there remain plenty of mortising, dadoing, and notching jobs for which chisels are ideally suited.

Whenever you pick up a chisel, keep your hands behind the cutting face of the blade and work the chisel away from your body. And since it takes both hands to operate a chisel, always clamp or anchor the material you're cutting. Finally, save yourself wasted effort and ruined materials by keeping your chisels sharp (see the tips on page 67).

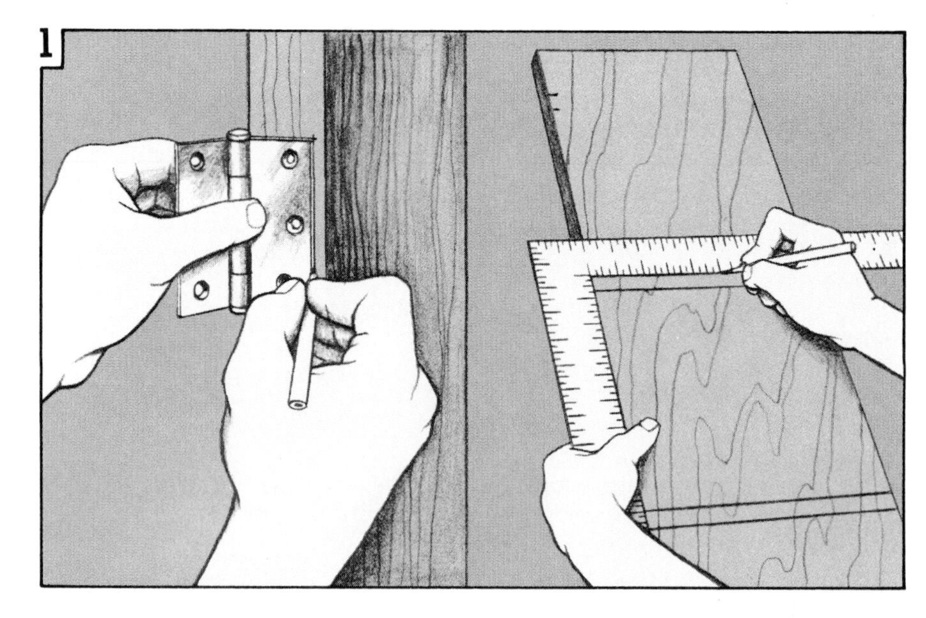

1 All of your care in executing a chisel cut is wasted unless you've marked accurately for the cut to begin with. When preparing to cut a mortise for a butt hinge, for example, use the actual hinge as your template. Position it correctly on the edge of the door and mark its perimeter with a sharp pencil. Also be sure to mark the "open" end of the mortise with the thickness of the hinge to indicate the depth of the cut.

When laying out dado cuts across the width of shelf uprights, marking square and straight are your watchwords. Hold a carpenter's framing square against a factory edge of the upright to square your marks, and outline the top and bottom of the cut.

2 To prevent the wood grain from splintering at the edges of chisel cuts, break the surface fibers along your cutting outline by scoring the wood with a simple utility knife to a depth of approximately ⅛ inch.

For deeper dado cuts, bring your circular saw into play. Adjust the blade to the depth of your dado and cut against a clamp-on guide to ensure that you stay on your outline mark.

3 Now hold the chisel with the beveled face of its blade facing the direction of your cut and at the angle shown. Always cut in the direction in which the wood's grain runs to the edge of the board. If you don't, the chisel can follow, descending the grain lines,

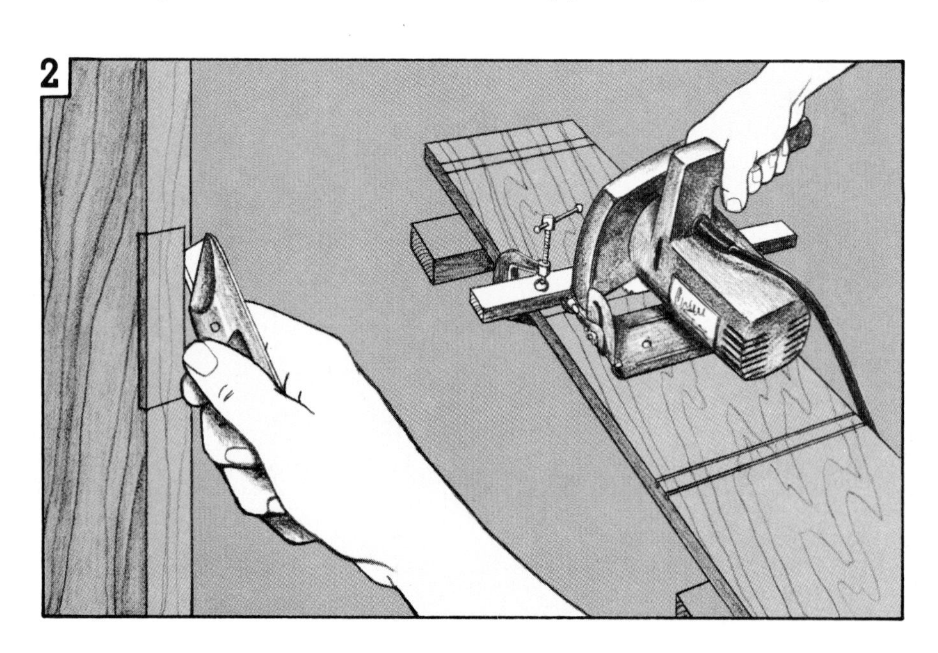

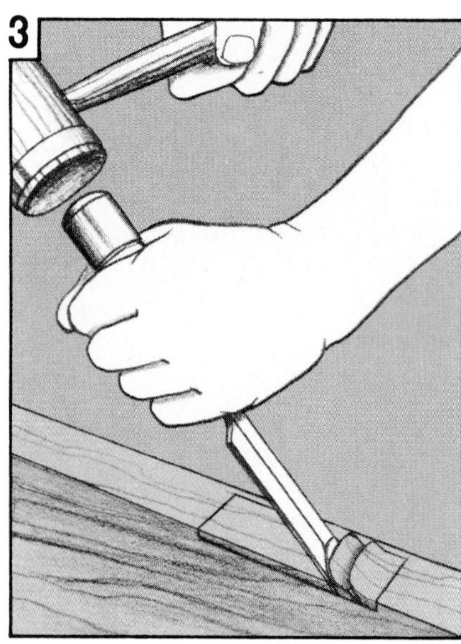

and gouge deeper than you intend. With your free hand, use a wooden-headed mallet to drive the chisel to the depth of the mortise, making several "notches" across its width.

4 To complete a dado cut after sawing to depth along its top and bottom, chisel out the wood between the cuts. Firmly tap the chisel with the palm of your hand. Start at an edge and hold the chisel at a slight upward angle as shown.

After you've removed a wedge of wood that extends to the center of the dado, do the same from the opposite edge. Now with the chisel bevel still face up carefully remove the remaining wood, guiding the chisel with your hand wrapped around the blade shank as shown.

5 It's hard to remove wood for a deep mortise cut with only a chisel. Even when there's enough handle clearance to maneuver the blade to the bottom of the mortise, you run a great risk of splitting the wood.

In situations like this, your cut will be easier and more accurate if you begin by drilling a series of overlapping holes within the scored outline of your mortise. If possible, choose a bit diameter the same size as the width of your mortise. To avoid drilling too deep, use the improvised depth gauge shown on page 43.

Finish the cut with a chisel, driving it with a mallet if necessary. Remember to hold the chisel with its beveled face toward the inside of the mortise.

6 Cutting a hefty notch from a 2x4 would be hard work for an unaided chisel, but becomes considerably easier with the help of a circular saw.

After marking for the notch, including its depth, adjust your saw blade to cut to this depth. Now make cuts at the top and bottom of the notch, and fill in with closely spaced cuts across the entire width of the notch.

Finish the cut by positioning the chisel blade at your depth-of-cut mark with the beveled edge facing toward the face of the notch and driving it with a mallet. Again, begin cutting at a slight upward angle that becomes more horizontal as you chip away more of the wood.

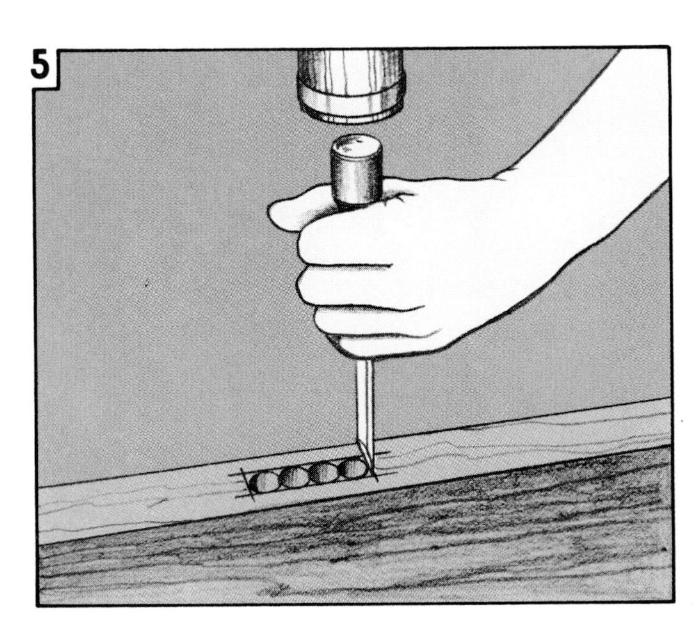

Drilling Techniques

Things were different before the advent of the electric drill. If a carpenter wanted to bore a small hole into or through a material, he'd reach for a hand drill or a push drill. And for larger holes, he'd pull his trusty brace down off the tool rack and fit it with the appropriate bit for the job.

But the electric drill has changed all that. This one tool has, to a great extent, made all other boring tools obsolete. Just insert the correct bit in its chuck (see page 12 for some of your bit options), and it will do any job any of the other drills can, but a great deal faster. Following are some techniques you should find helpful as you use this jack-of-all-trades boring machine.

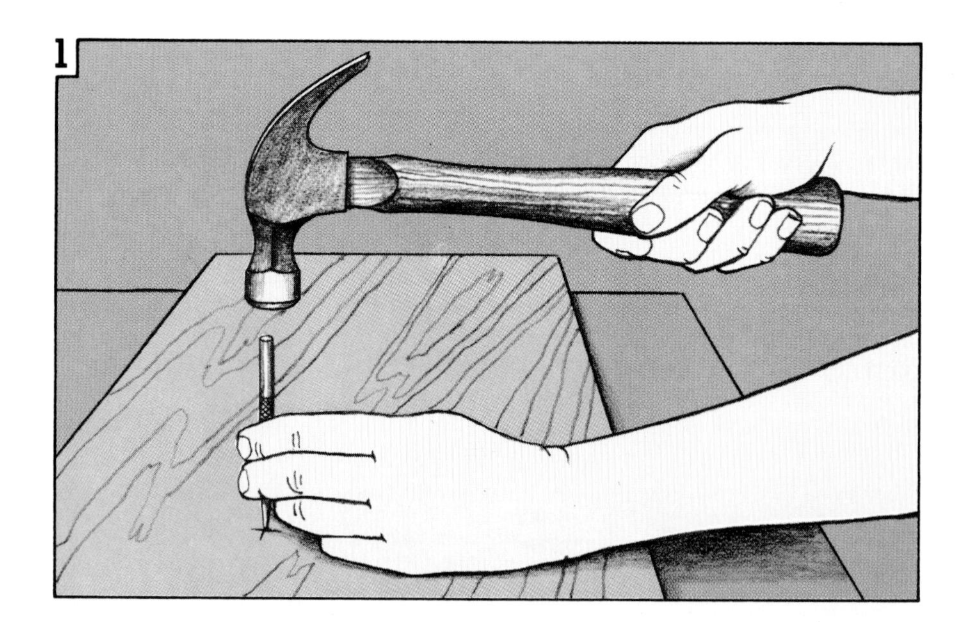

1 Because drill bits cut using a rotary motion, they tend to skate away from their intended location when you begin boring holes. To prevent this (and keep the bit from marring the surface of your work), make a shallow pilot hole with an awl or some other sharp-pointed object. In softwoods, a gentle tap with the palm of your hand will do the job. But with hardwoods, you'll probably need to tap the awl with a hammer.

With this done, grip the drill with both hands and center the bit in the pilot hole. Apply firm pressure and begin drilling. If you have a variable-speed drill, bring it up to full power gradually.

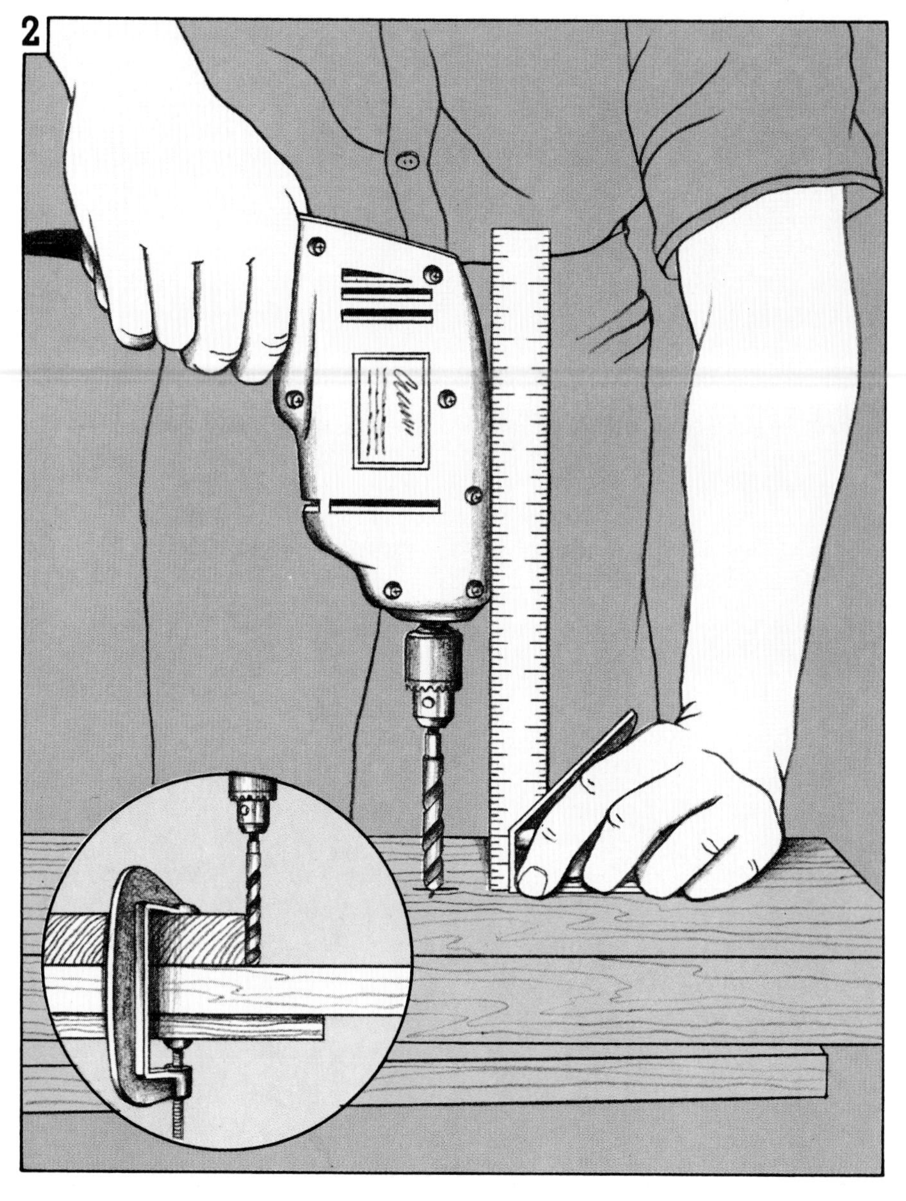

2 In most instances, you'll want the holes you drill to be perpendicular to the work. Though you can guarantee this by buying a drill press or an attachment that approximates a small drill press, you can achieve good results—less expen-

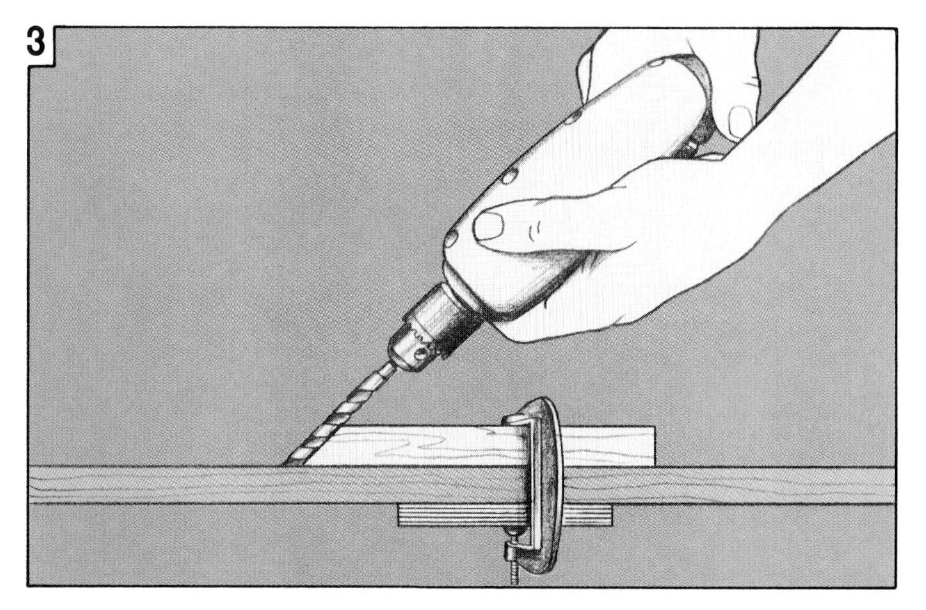

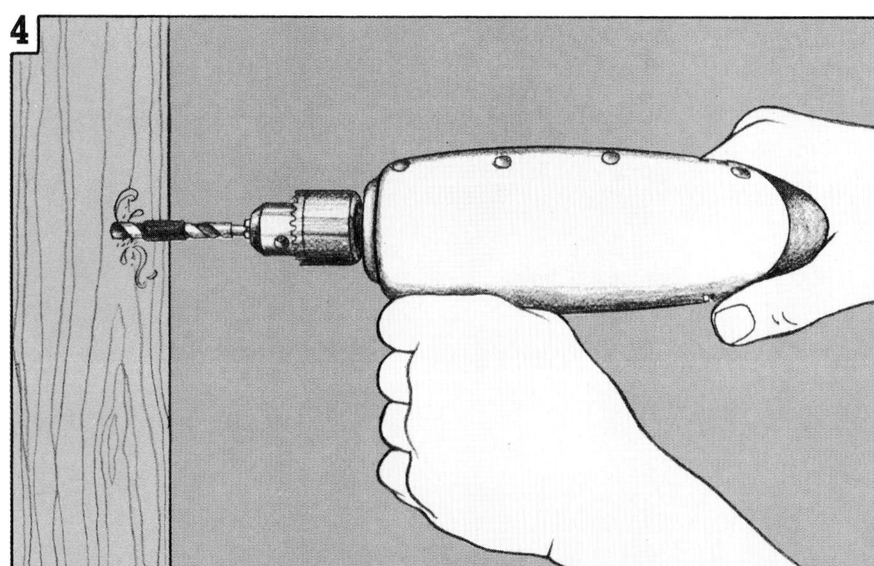

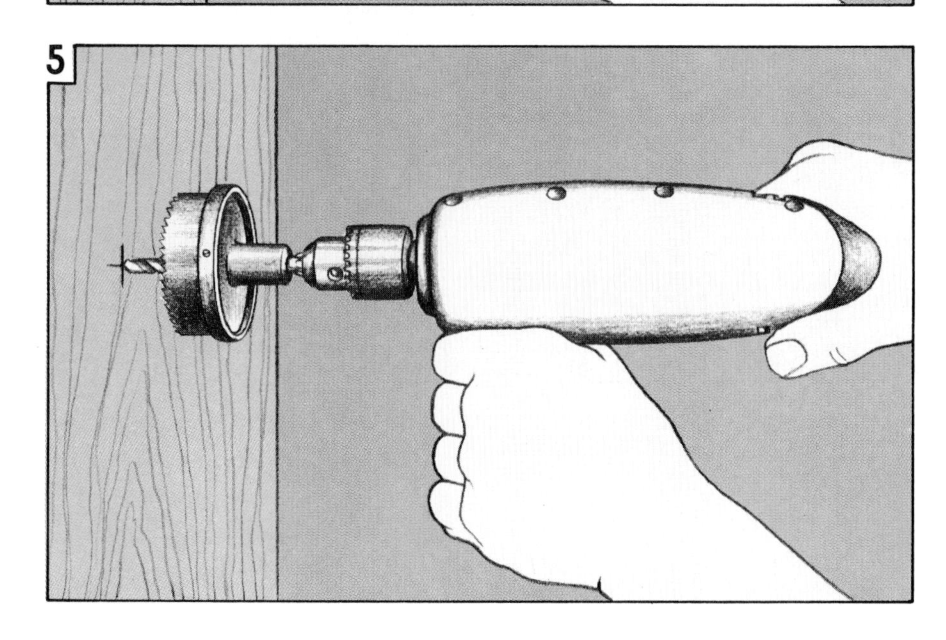

sively—by checking the bit for square as it enters the material. To do this, hold the body of a combination square (or the handle of a try square) firmly against your material, aligning the drill against the blade as shown. If your drill has a tapered body, you can check the bit for square by placing the edge of a piece of scrap lumber next to the bit.

3 Sometimes you'll *want* your bit to enter the material at an angle, in which case maintaining that angle becomes the real trick. Fashioning a jig will help simplify things a lot. Begin by cutting the edge of a piece of scrap lumber to the desired angle of your hole (see page 36 for how to make bevel cuts). Now C-clamp the scrap to the surface of your material so it aligns the tip of the bit exactly on your center mark. Carefully guide the bit into the material, keeping its sides in contact with the edge of the guide.

For steep angles, angle your pilot hole to prevent the bit from skating out when you begin drilling the hole.

4 For situations in which you want to drill one or more holes to a certain depth, simply wrap masking or electrical tape around your drill bit so the bottom edge of the tape contacts the surface of the material at the desired depth. Drill with gentle pressure, and carefully back the bit out of the hole as soon as the tape touches the material's surface.

5 When using a hole saw to bore large-diameter holes, make a pilot hole on your center mark to guide the starter bit. And to ensure that the underside of your material doesn't splinter when the bit penetrates it, place a piece of scrap stock underneath. Here again, keep your drill perpendicular to the material. (*continued*)

Drilling Techniques *(continued)*

6 Another technique for preventing the back side of your material from splintering involves drilling through the material until just the tip of the bit penetrates the back side. Then carefully back the bit out of the hole and finish drilling from the other side, using the pilot hole you've just made.

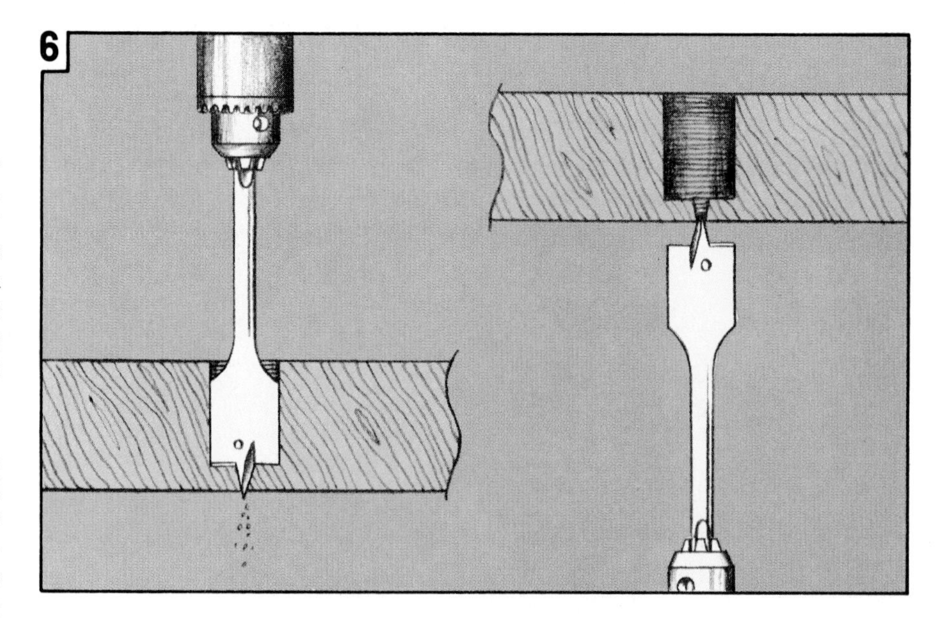

7 When you drill deep holes in thick material, wood chips will build up in the hole, clogging the bit and causing it to bind. To minimize this problem, feed the bit into the wood slowly, and back the bit out of the hole frequently with the drill's motor still running. This brings the trapped wood particles to the surface.

If you're working with extremely sappy wood, shavings may clog the bit's flutes. If this happens, let the motor stop and then use the tip of a nail to scrape them out.

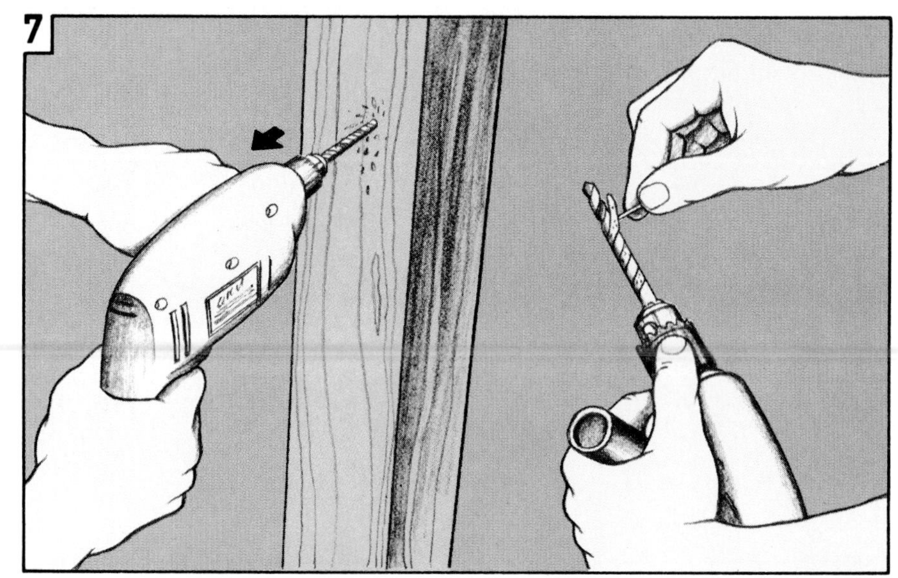

8 When you use wood screws to fasten two pieces of material together, especially hardwoods, always provide adequate clearance for the screw, to ensure easy driving and to avoid splits. Here's one way to proceed.

First, drill a hole partway through the top material to accommodate the screw shank. Next, drill a deeper pilot hole for the screw threads, using a smaller-diameter bit. And if you plan to *countersink* the screwhead (see page 49 for more on this), use a countersink bit to make a shallow depression at the top of the shank clearance hole.

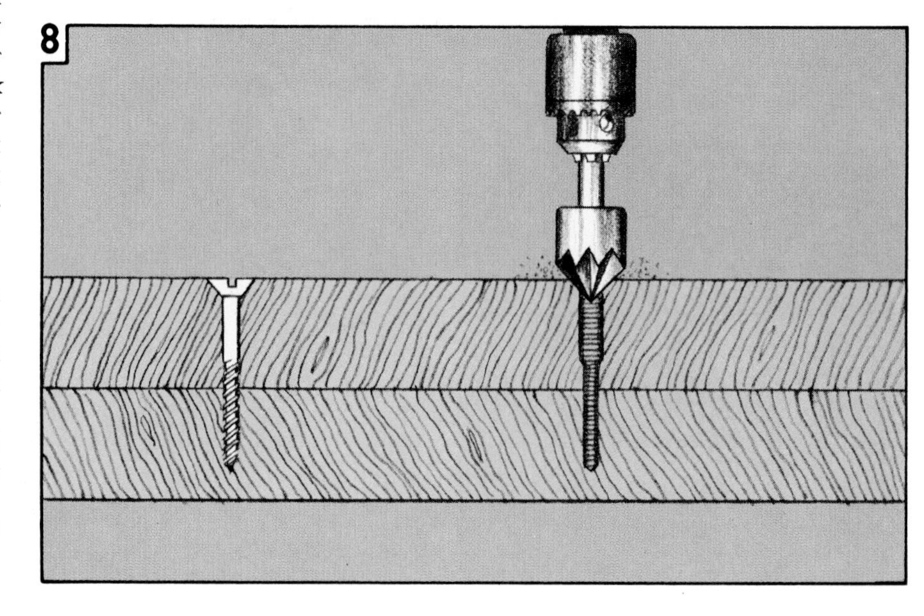

9 Though the technique discussed for sketch 8 will work, it is time-consuming. With a combination countersink/counterbore bit, you can cut your work in half. Available in sizes for most common wood screws, these bits prepare the way for screws in one simple operation. Just select the proper size bit for the screw you'll be driving, and drill the hole. If you want the screwhead flush with the surface, drill to the top of the bit's scored area. To counterbore the screwhead, drill deeper.

10 Measuring and marking for a hole often can be more time-consuming than the actual drilling. So if a project calls for drilling holes in exactly the same location in several pieces of lumber, try this highly accurate, work-reducing technique. Align the edges of your pieces, stack them in a bundle no thicker than the length of your drill bit, and C-clamp them together. Now center-mark the top surface, make a pilot hole, and drill through all the pieces. Use a piece of scrap stock beneath your stack or finish your hole from the underside to avoid splintering the bottom piece.

11 If the job calls for many holes, you can save a great deal of time by building a jig. With one of these, once you've done the initial setup, you can drill any number of accurately placed holes as fast as you can feed the material into position. The jig's bottom platform prevents splintering.

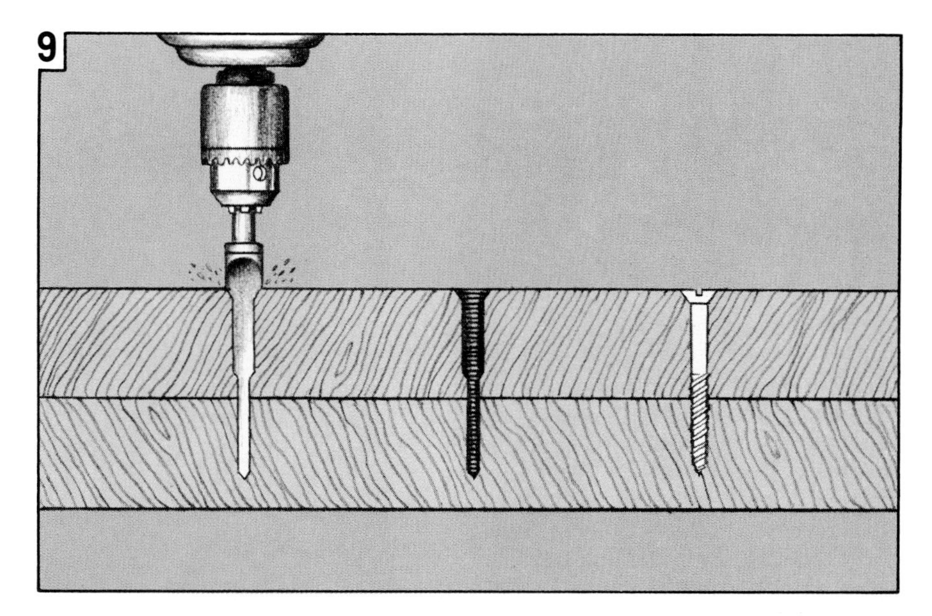

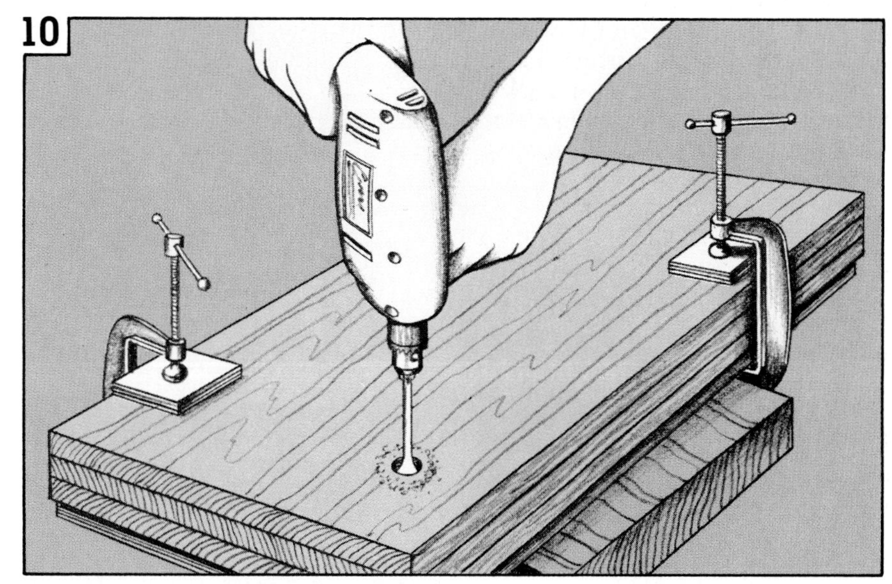

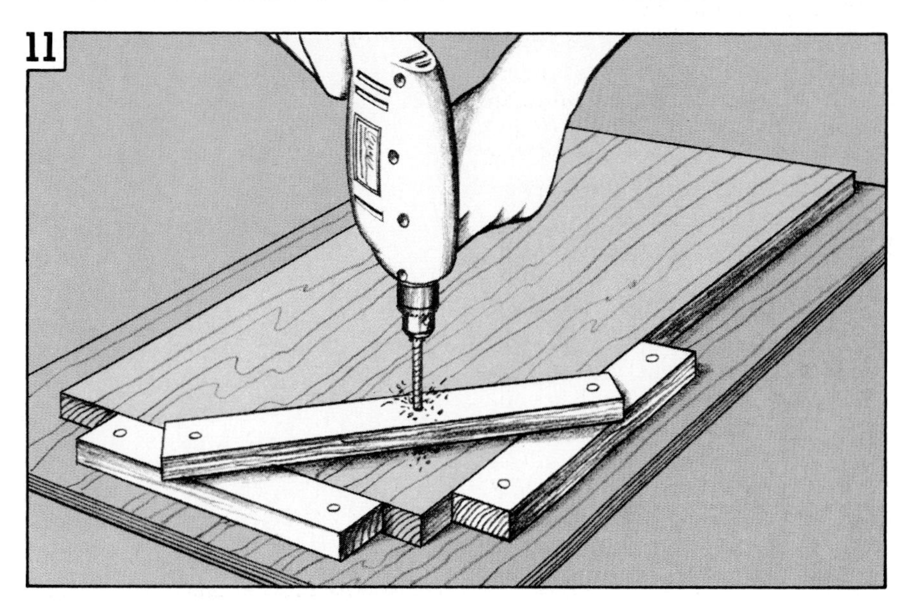

Fastening Techniques

How To Drive and Pull Nails

If you've ever watched an experienced carpenter wield a hammer, you've probably marveled at the speed he can fasten together the components of a large project. Impressive, yes, but for your purposes, speed isn't what counts. What does is your ability to drive and remove nails without hurting yourself or damaging the materials. And to do that requires only two things: a knowledge of a few basic techniques, and *plenty* of practice.

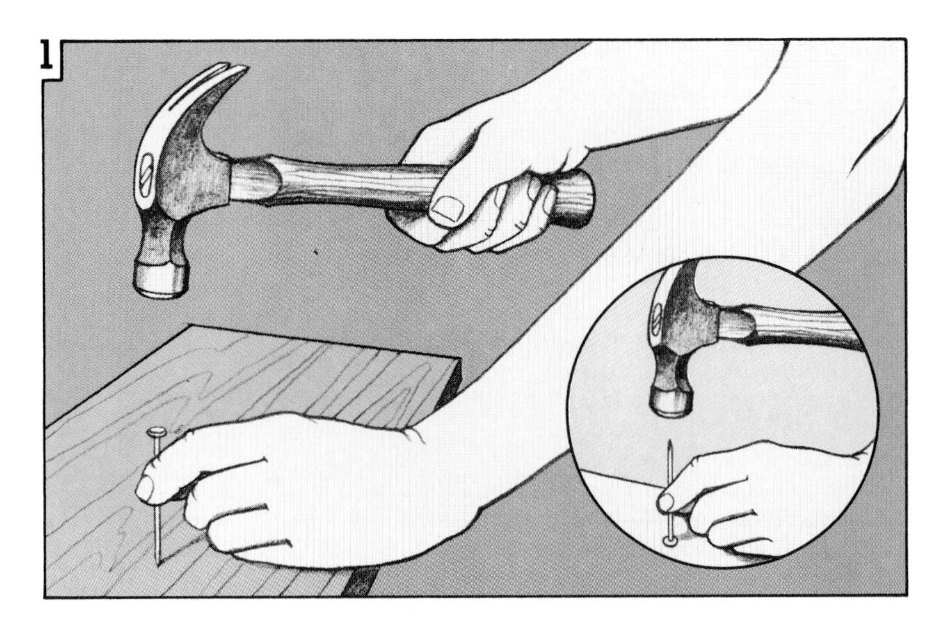

1 To make sure that the hammer strikes the nail—not your fingers—and that the hammer's blow will drive the nail squarely into the work, grasp the nail near its head and the hammer near the end of the handle. *Keep your eye on the nail* as you swing the hammer downward, and let the weight of the hammer's head do the driving. If you must drive a nail near the end of a material (especially lumber), blunt its point with a hammer. The blunted point will punch through the wood fibers and thereby reduce the chance of splitting the material.

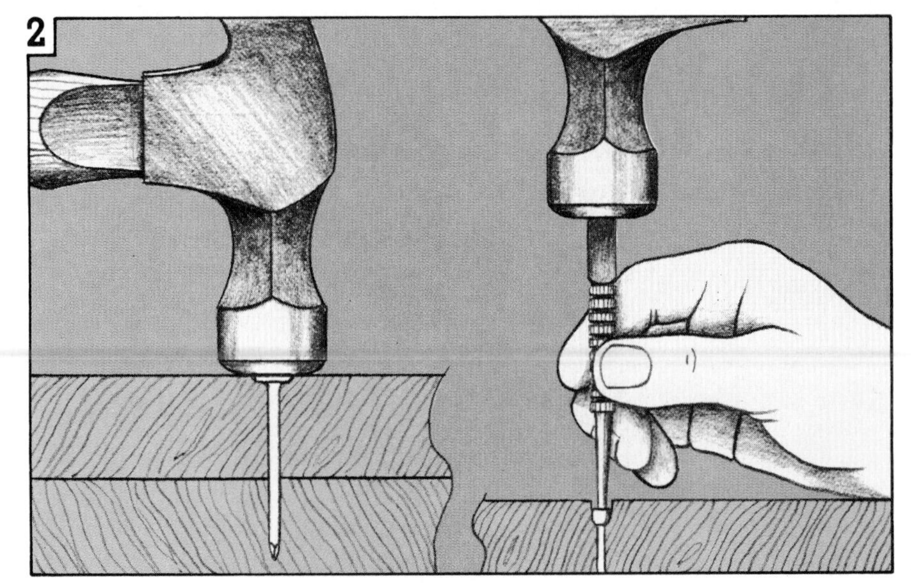

2 The last blow from the hammer should push the head of most nails flush with the surface of the wood. (The convex shape of the hammer's face should allow you to do this without marring the surface of the material.) With finishing or casing nails, though, you'll want to drive the heads below the surface. Use a nail set and fill the hole with wood putty later.

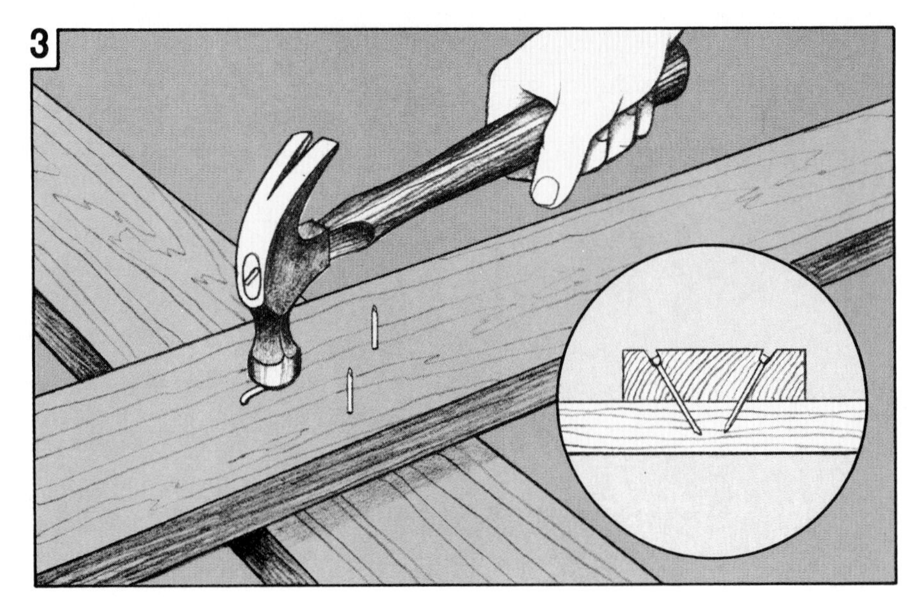

3 For maximum holding power in rough projects, use nails about an inch longer than the thickness of the pieces you're fastening. Drive the nails, then turn the members over and *clinch* (bend) the exposed portion of the nails in the direction of the grain so they're nearly flush with the surface. For showier projects or in situations where you can't get at the back side of the material, drive pairs of nails in at an angle.

4 For best results when fastening a thin sheet of material to thicker stock, nail through the thinner one into the thicker.

5 When driving several nails along the length of a board, stagger them so you don't split the board.

6 With some materials, especially hardwood lumber and moldings, you'll have to drill pilot holes before driving the nails to avoid splitting. Make the pilot holes slightly smaller than the diameter of the nails.

7 From time to time, you'll strike a glancing blow or encounter a knot while driving a nail, both of which may bend the nail. When this happens, remove the deformed nail and start again with a new one. To keep from marring the surface of the material and for better leverage, place a piece of scrap material beside the nail, slip the hammer's claw beneath the nailhead, and pull back on the handle. With long nails, you may have to substitute a thicker piece of scrap once you get the nail partway out.

8 How you remove nails that have been driven flush with the surface depends on the situation. If you're lucky enough to have access to the back side of the joined materials, first strike the joint (*continued*)

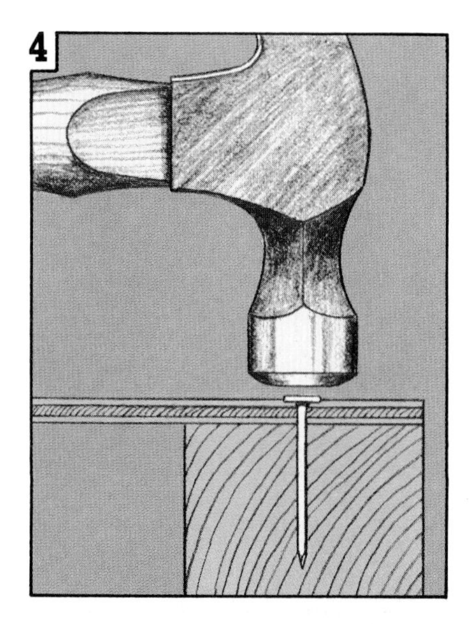

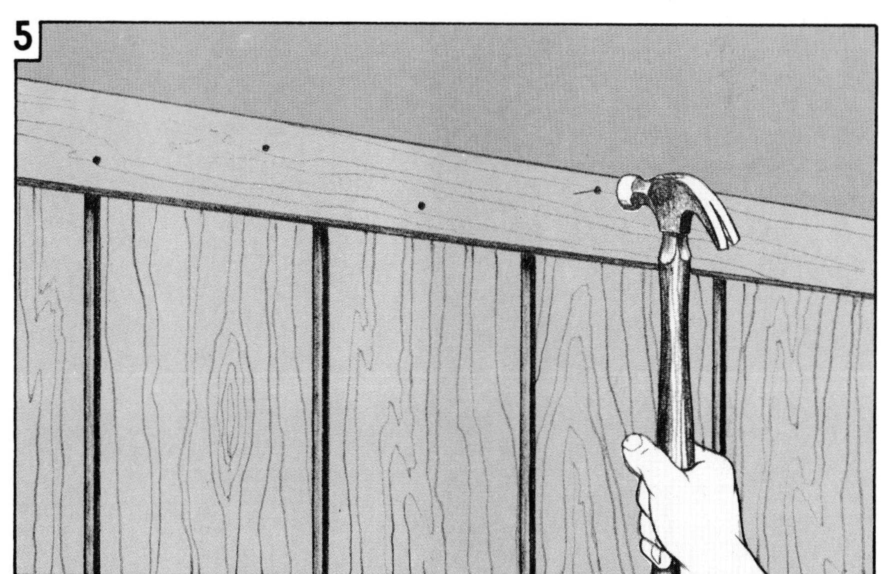

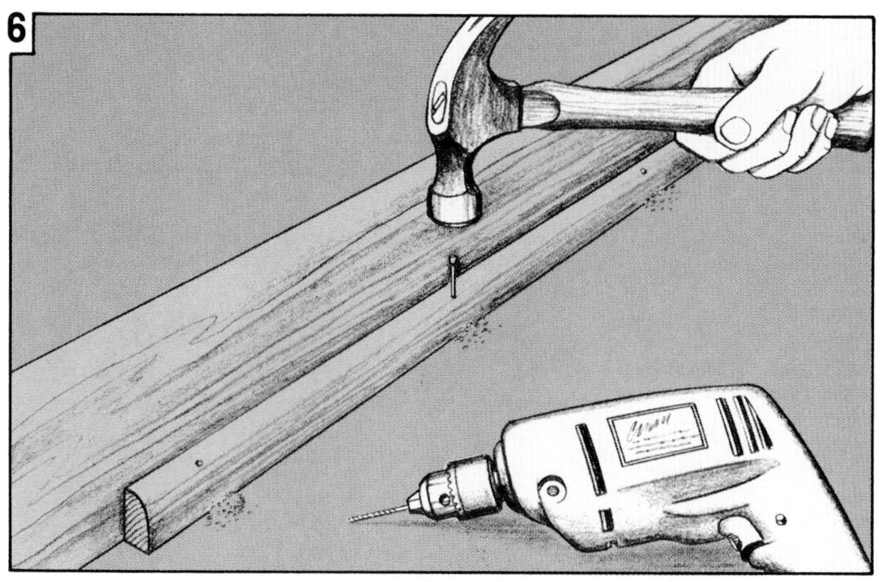

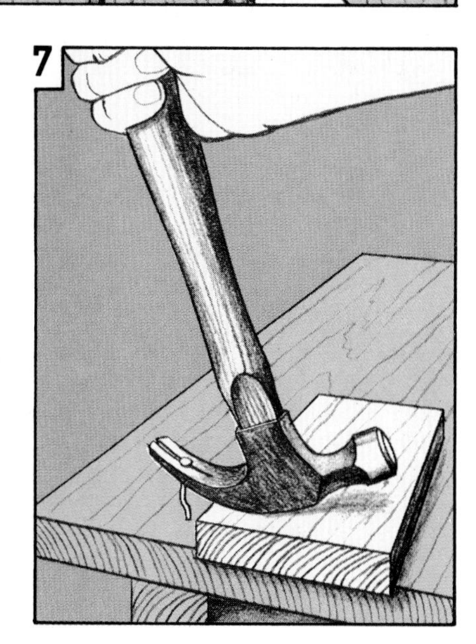

How To Drive and Pull Nails *(continued)*

from behind, then hammer the members back together again from the front side. This should pop the nailheads out far enough for you to get hold of them with a hammer.

9 No access from behind? Then try tapping the V-shaped claw of a pry bar (it's thinner than the claw of a hammer) underneath the nail's head. Then pry the nail up far enough so you can get the head of a hammer claw underneath the nailhead. This technique will leave its mark on the surface of the wood, but often you have no other option.

10 Sometimes, as in this situation, you can "disjoin" two nailed-together members by sawing through the nails. Then when you have separated the members, pound the nails from the back side with a hammer and a nail set to gain access to the nailheads.

11 Removing moldings intact is a difficult chore that requires care and patience. Because the nails holding them in place have been sunk and the cavities filled, your best bet is to first drive the nails deeper into the stock, using a hammer and nail set. Then slip or force a prybar under the molding near each nail, and rock the bar back and forth to free the molding.

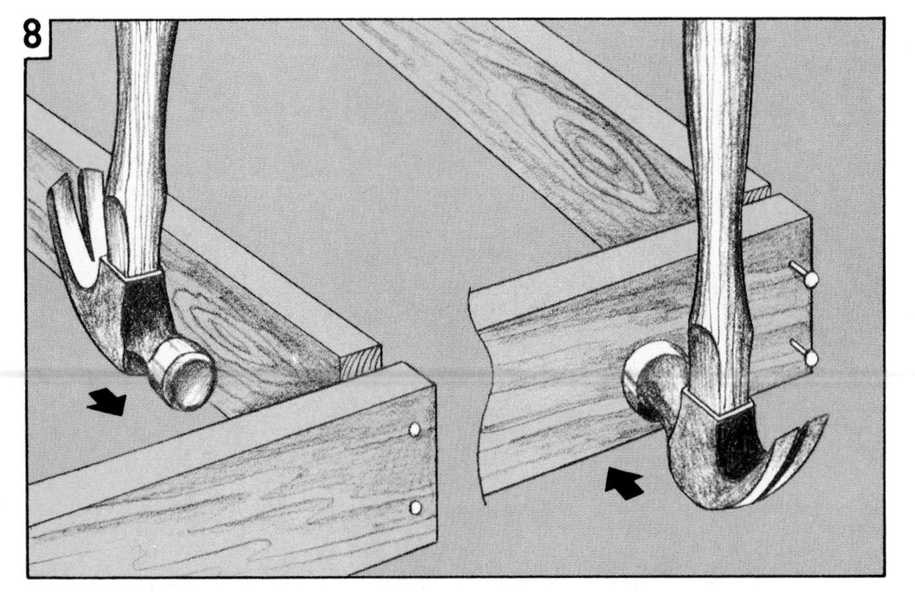

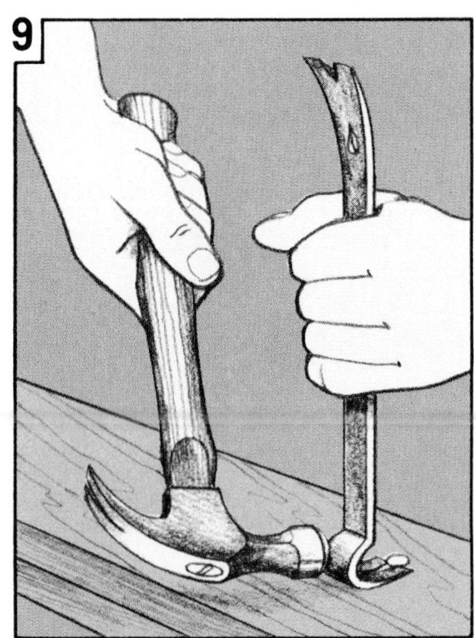

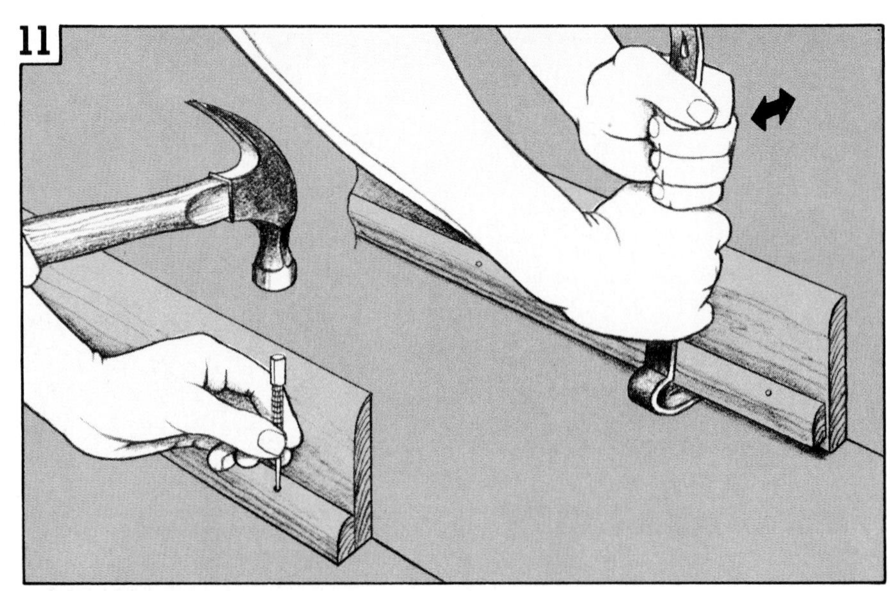

How To Drive and Remove Screws

It's not difficult to see why screws do such a good job of fastening materials. The many threads along the shank of every wood screw grip the wood fibers of the members being joined and, when the screw has been driven home, exert tremendous pressure against the screwhead.

Unfortunately, the same characteristics that make screws the super fastener they are can cause you problems when driving and removing them. Here, we tell you how to minimize such hassles.

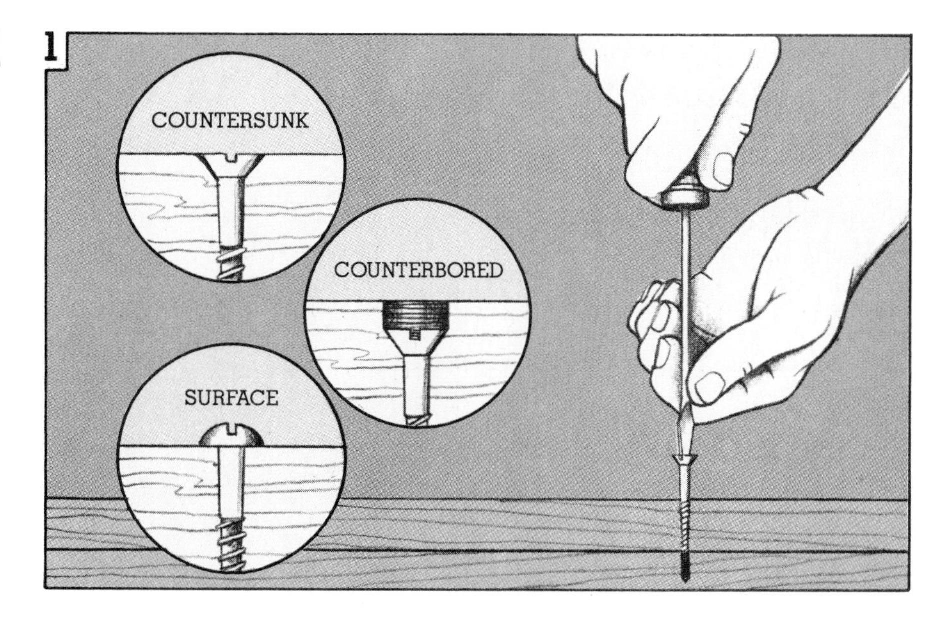

1 Once you've decided on where a screw will go, drill an appropriately sized pilot hole in one of the two ways shown on page 44. (Note that depending on the need, the screwhead can rest on top of, flush with, or below the surface.) Then select the correct screwdriver for the screw you're driving (it should fill the slot completely for maximum torque). Set the screw in the pilot hole, and with one hand on the screwdriver's handle and the other steadying its blade in the screw's slot, slowly turn the screwdriver clockwise. Apply only moderate pressure at first.

2 If the going gets tough, exert pressure on the screwdriver with the palm of one hand and turn it with the other. And if you still can't drive the screw all the way home, remove it and drill a slightly larger pilot hole or lubricate the threads with candle wax or soap and try again.

3 Driving lots of screws (or even just a few large ones) by hand can tire you out in a hurry. To cut down on fatigue and to speed the whole operation, drive the screws with an electric drill and a screwdriver bit. Make sure the bit is firmly seated in the screw slot before beginning, and don't drive the screw too quickly. If you damage the screw's slot, you may be able to salvage the screw by deepening its slot with a hacksaw.

4 Faced with a hole that's too big for the screw you're driving? Insert a toothpick, a sliver of wood, or a bit of steel wool into the hole. Tightening down the screw will force the filler material against the wall of the hole and hold the screw tight. *(continued)*

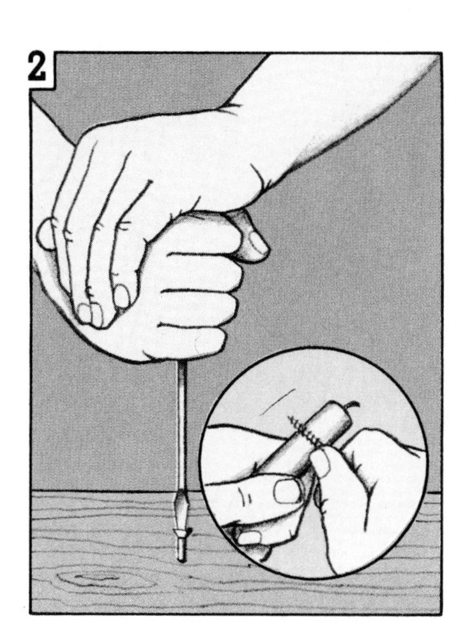

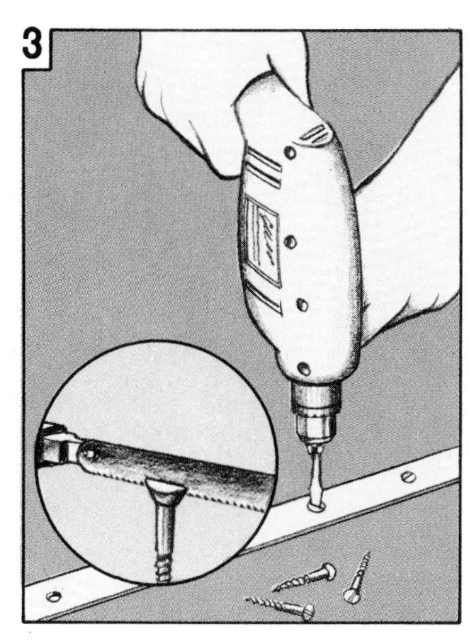

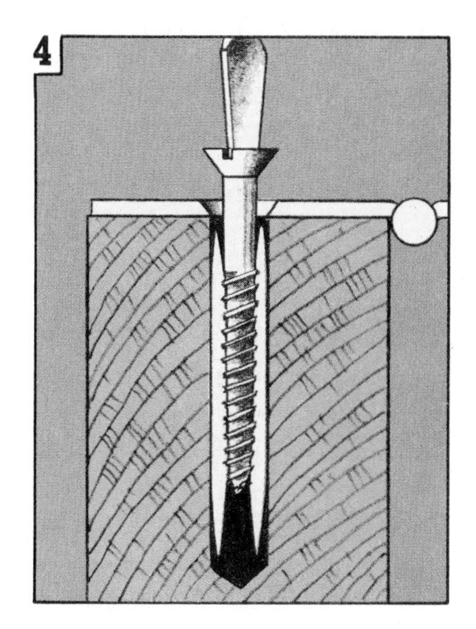

How To Drive and Remove Screws *(continued)*

5 Sometimes, screws do their job of holding materials together too well. To remove a balky one, first try working the screw back and forth. If you're lucky, that may be just enough to free the bind. Be careful that you don't damage the screwhead while attempting to loosen the screw. If you do and the screw is a straight slot type, you'll have to employ a screw extractor as shown in sketch 7. For Phillips head screws, drill a small hole in the base of the screwhead to give the screwdriver blade more surface to grab onto.

6 No luck? Then heat up the screwhead by holding a soldering gun to it, or strike it several times with a hammer and nailset. Both of these techniques have helped remove their share of frozen screws.

7 If what you need is greater turning power than your hands and the screwdriver handle can supply, team up a square-shank screwdriver and an adjustable-end wrench.

And if even that fails, drill a hole into the center of the screwhead, then thread in a screw extractor. As you turn this tool counterclockwise, it seats itself in the screwhead then (if you're fortunate) turns out the screw.

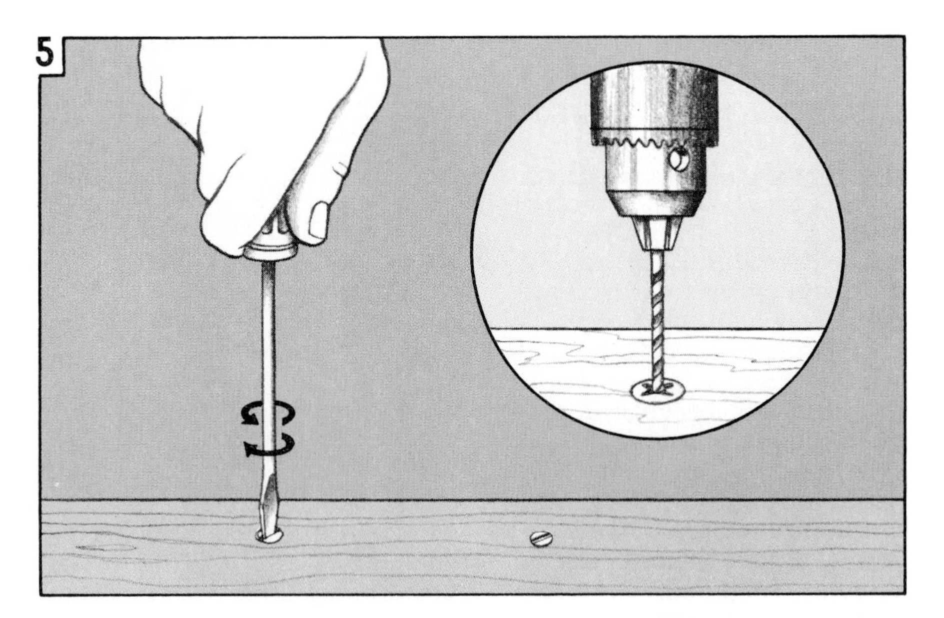

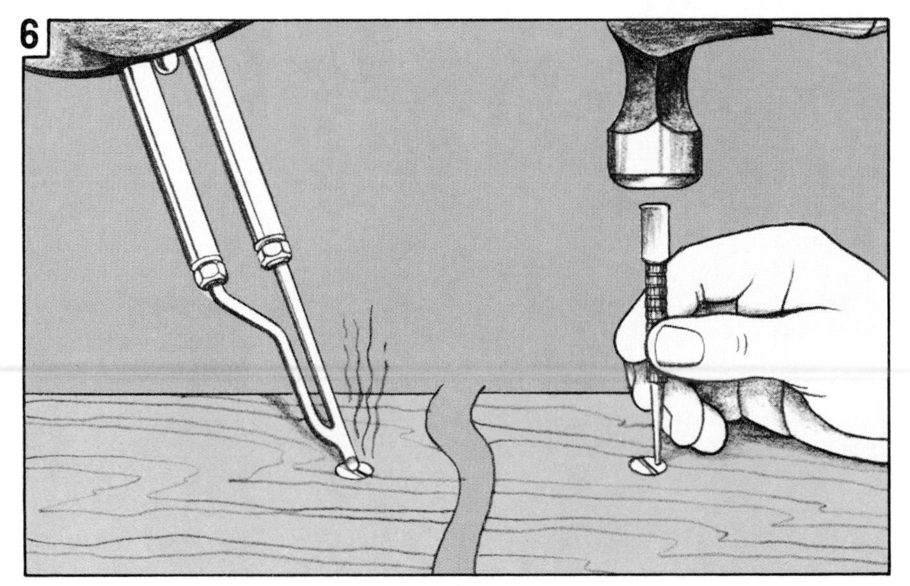

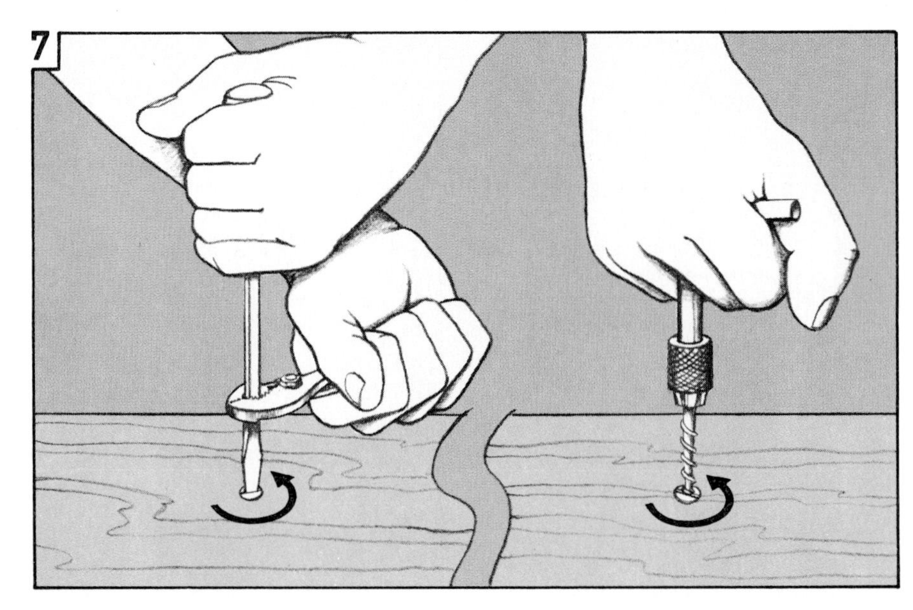

How To Tighten and Loosen Bolts

Nails and screws depend on friction between the fastener and the wood to do their job. Not so with bolts, however. When you tighten a nut on a bolt, you're actually "clamping" adjoining members together, and in the process creating the sturdiest of all joints.

On this and the following page we limit our discussion to techniques involving machine and carriage bolts—the heavyweights. For information about two specialty fasteners in this category—toggle bolts and hollow-wall anchors (the types you'd use to fasten things to hollow walls)—see pages 54-55.

To install any kind of bolt, you must first bore a hole through the material. The steps after that differ somewhat depending on the type of bolt you use.

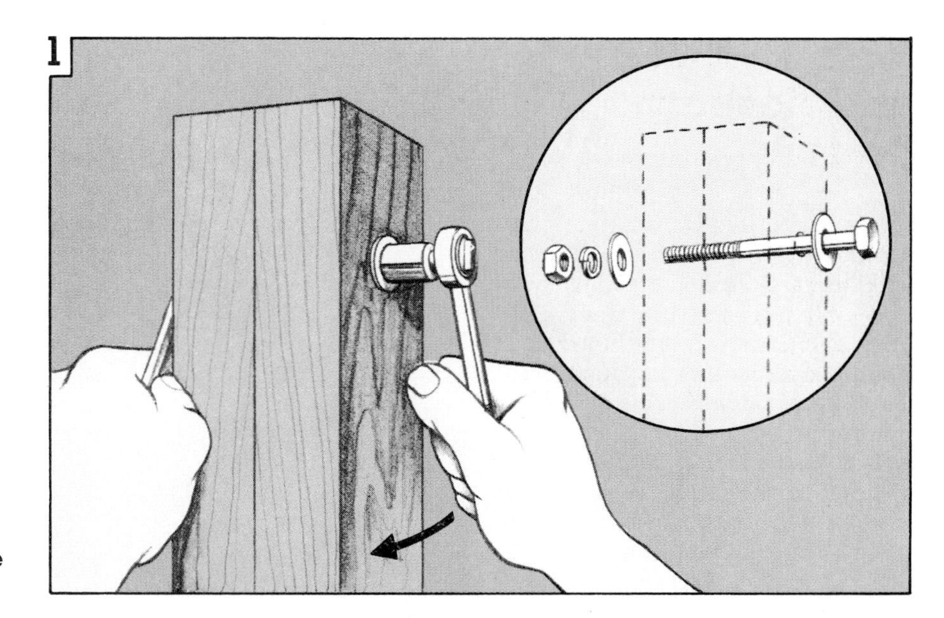

1 For machine bolts, which have threads running the length of the shank, first slip a flat washer onto the bolt, then slide the bolt through the hole. Put another flat washer, then a lock washer over the bolt, and follow these with a nut. (The flat washers keep the nut and the bolthead from biting into the wood; the lock washer prevents the nut from loosening.)

To draw the nut down onto the bolt, you'll need two wrenches; one to steady the nut, the other to turn the bolt's head.

2 To tighten down a machine bolt in hard-to-get-at places or when you have countersunk the bolt's head, you will need a socket wrench, possibly with an extension to reach into the recess. (Again, steady the nut by using a second wrench.)

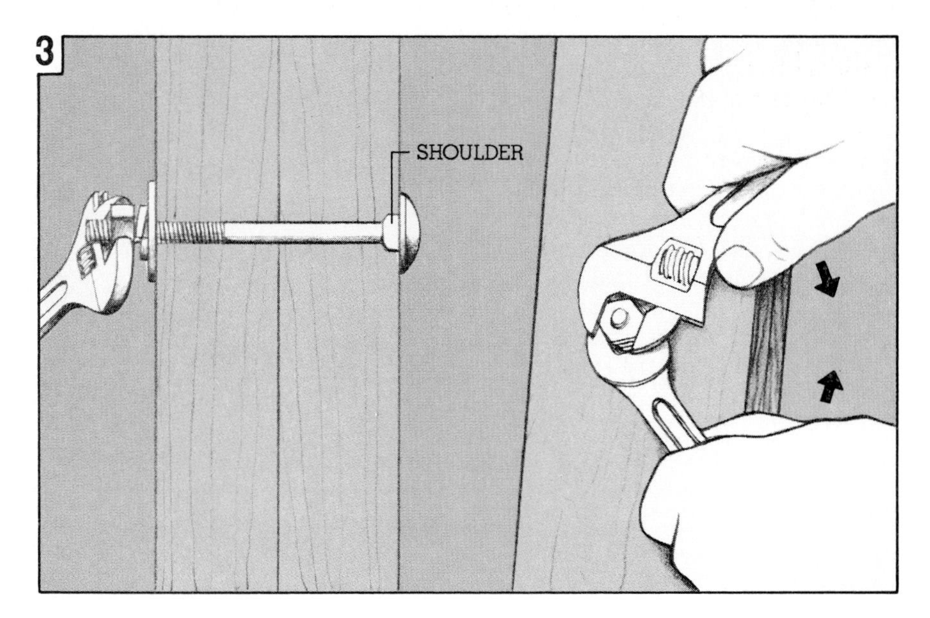

3 To tighten a carriage bolt, simply insert it into the hole (*continued*)

How To Tighten and Loosen Bolts *(continued)*

and tap its head flush with the surface. Then slip a flat washer, a lock washer, and a nut onto the bolt and turn the nut clockwise. The shoulder under the bolt's head keeps the bolt from spinning as the nut is tightened from the other side.

The lock washer should keep the bolt from working loose. But if you want to make absolutely sure of it, thread another nut onto the bolt, snug it up against the first nut, then "jam" the two together using the action depicted by the arrows.

4 Like any other fastener, a bolt can be difficult to remove, especially if it has been in place for quite some time. If the bolt's threads have rusted, scrub them briskly with a wire brush. And if that doesn't free things up, lubricate the bolt and nut with a few drops of penetrating oil.

5 Rust isn't the only cause of difficult nut removal. Damaged bolt threads can make the task equally hard. If you run into this situation, simply cut off the damaged part of the bolt with a hacksaw, as shown in the drawing. Note the angle of the saw blade. You can adjust many hacksaws to cut at a 90° angle to the material.

6 You also may find that the bolt itself is stuck in the hole. To remedy this problem, use another bolt as a punch. A few taps from the back side with a hammer should back the bolt out partway. Then grip the bolt's head with a pair of pliers and wiggle and twist it the rest of the way out.

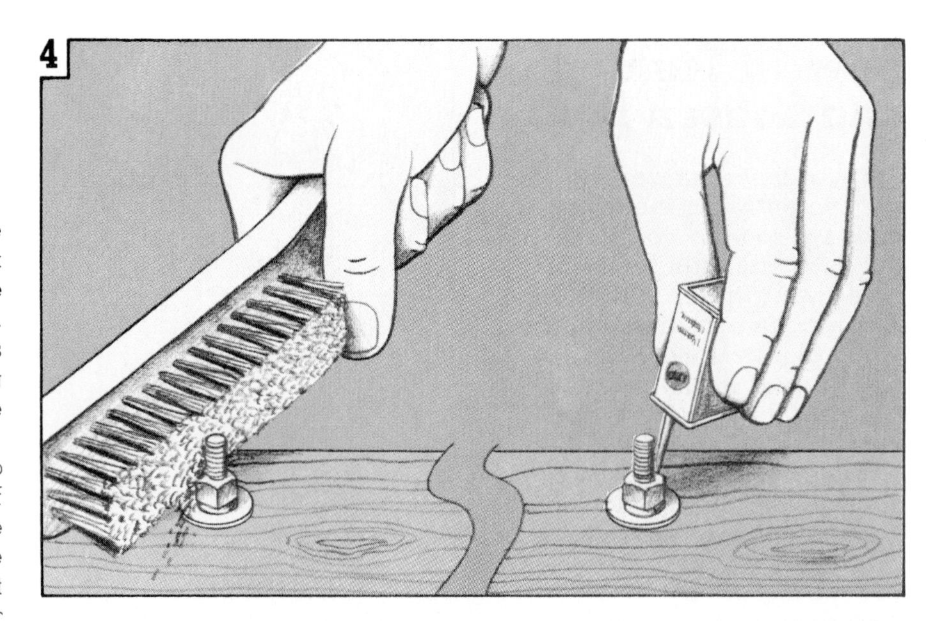

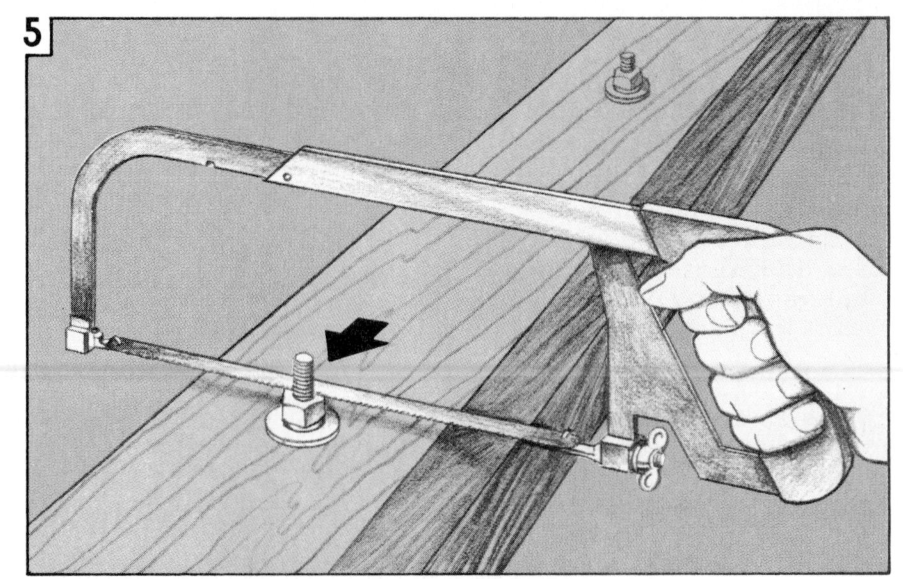

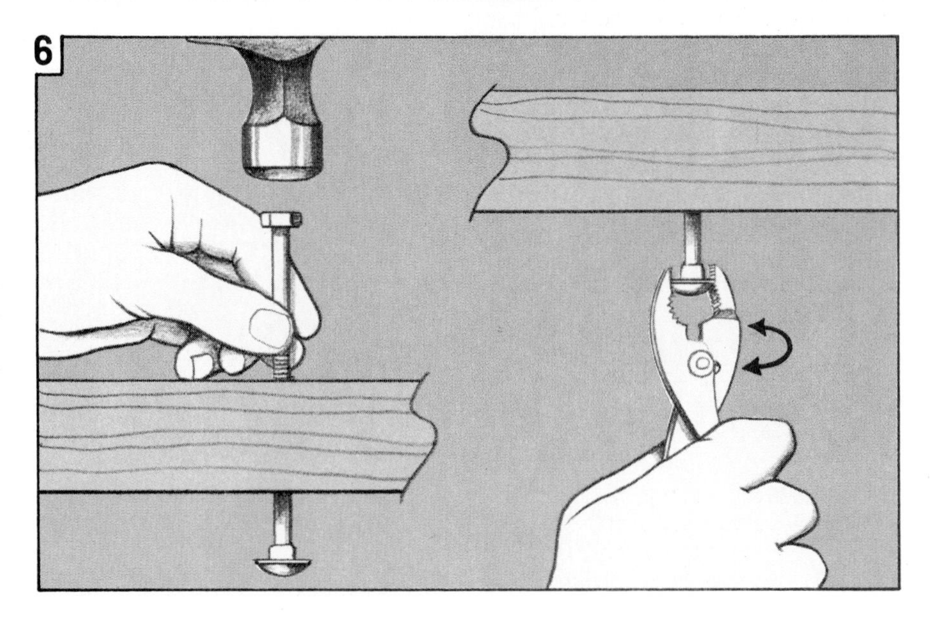

Using Glues and Adhesives

Today's glues and adhesives are incredibly strong, easy to work with, and durable. So it's no wonder that amateur and professional carpenters alike depend on these "super" fasteners for so many of their needs. The two types discussed here—wood glues and panel adhesives—involve different application techniques, both of which we explain.

1 For best results with wood glues, first check the members to be joined for a snug fit. Then apply a thin, even coat to both surfaces.

2 Now fit the members together and secure the joint with clamps or a weight until the glue sets up. Keep the following in mind when using clamps.

- Protect the good surfaces by placing scrap material between the clamp jaws and the project.
- Use as many clamps as you need to make sure that the glue-coated surfaces will remain in contact.
- Apply just enough clamping pressure to create a tight seal without distorting the wood.
- Recheck the fit after tightening the clamps, and make any necessary adjustments.
- Wipe off any glue that flows from between the clamped members. (A few tiny droplets of glue along the edges of the seam is a good sign that you've used enough but not too much.)

3 Unlike wood glues, panel adhesives don't require clamping. Instead, you follow this four-step procedure. (1) Lay a bead of adhesive on one of the surfaces to be joined. (2) Press that surface against the mating surface. (3) Pull the materials apart slightly. (4) And after waiting the time specified on the adhesive tube, press the materials back together again.

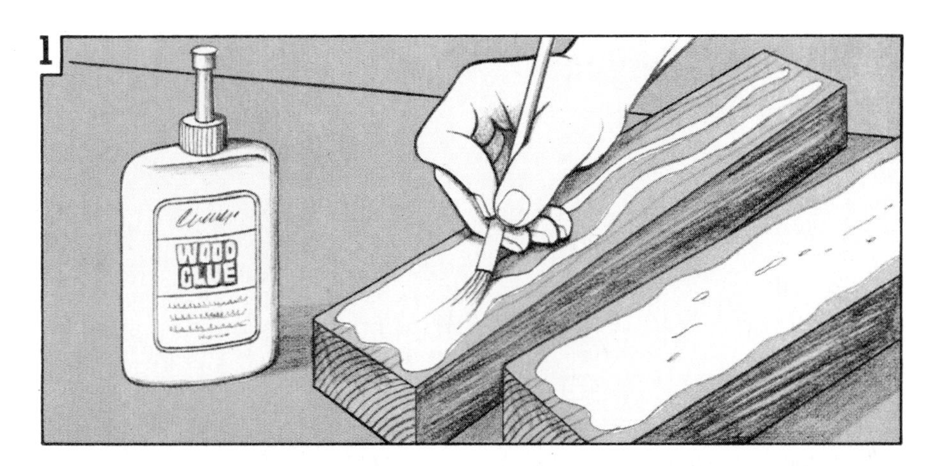

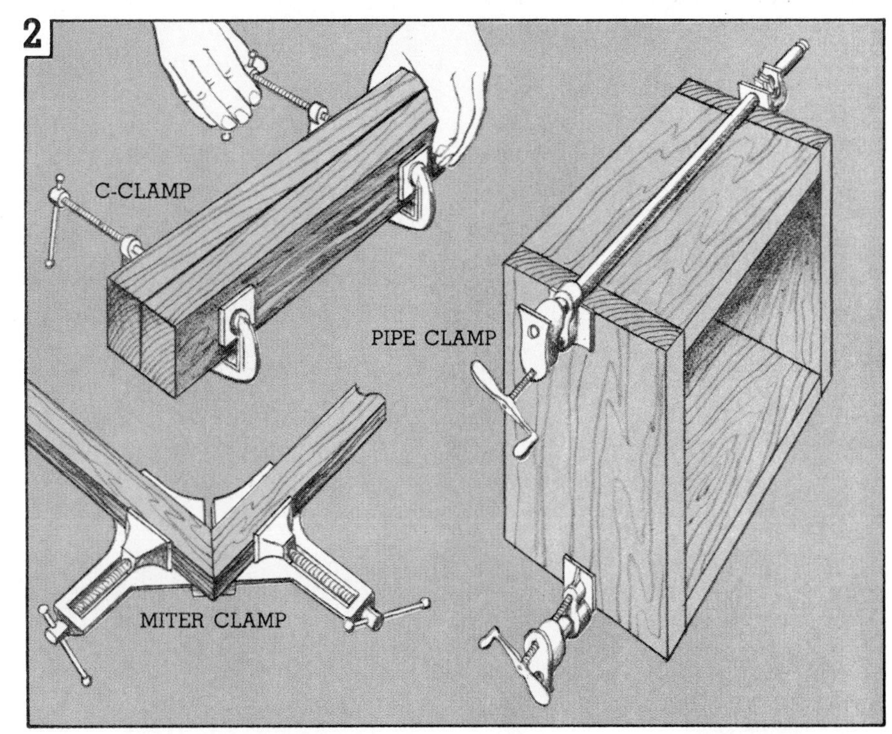

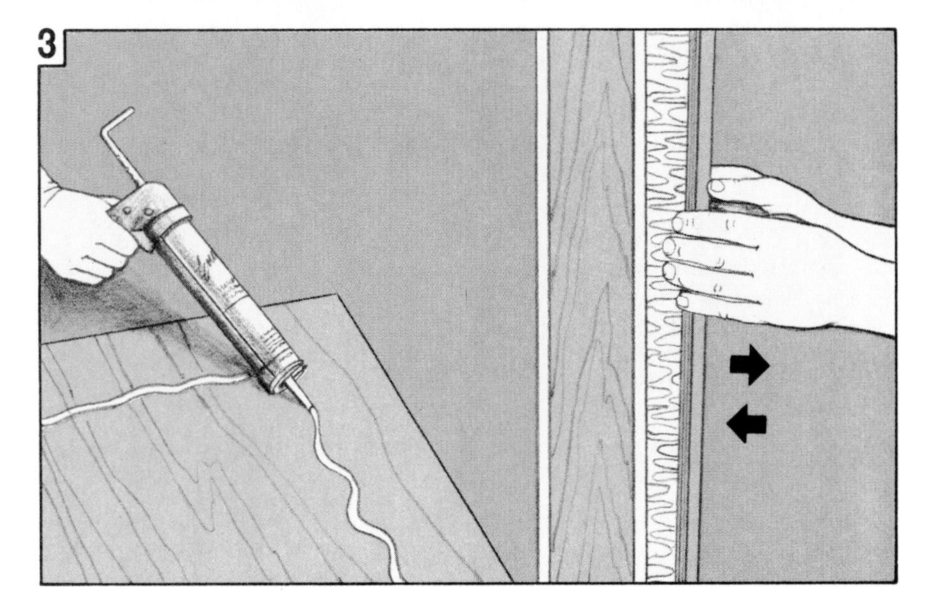

Making Simple, Strong Joints

Strong, good-looking wood joints have long been the hallmark of skilled carpenters and cabinetmakers. That's because these pros realize how important good joinery is to the end product.

Here, we show you several different ways to successfully join two pieces of wood at an angle, none of which requires cabinetmaker expertise. Before you make your joints, though, be sure to review the material beginning on page 46. It tells you how to work with the fasteners you'll be using.

To form a *butt joint*, position the end of one member against the face or edge of another member. Though not one of the strongest in and of itself, when reinforced by any one of the methods shown here, butt joints yield satisfactory results for many projects in which appearance isn't a prime factor.

Lap joints offer greater strength than butt joints and, at least in the second and third examples shown, better looks. To make an *overlap* joint, simply lay one of the members atop the other and nail or screw it in place with at least two screws or nails driven as shown.

For a *full-lap* joint, cut a recess into one member that's as deep as

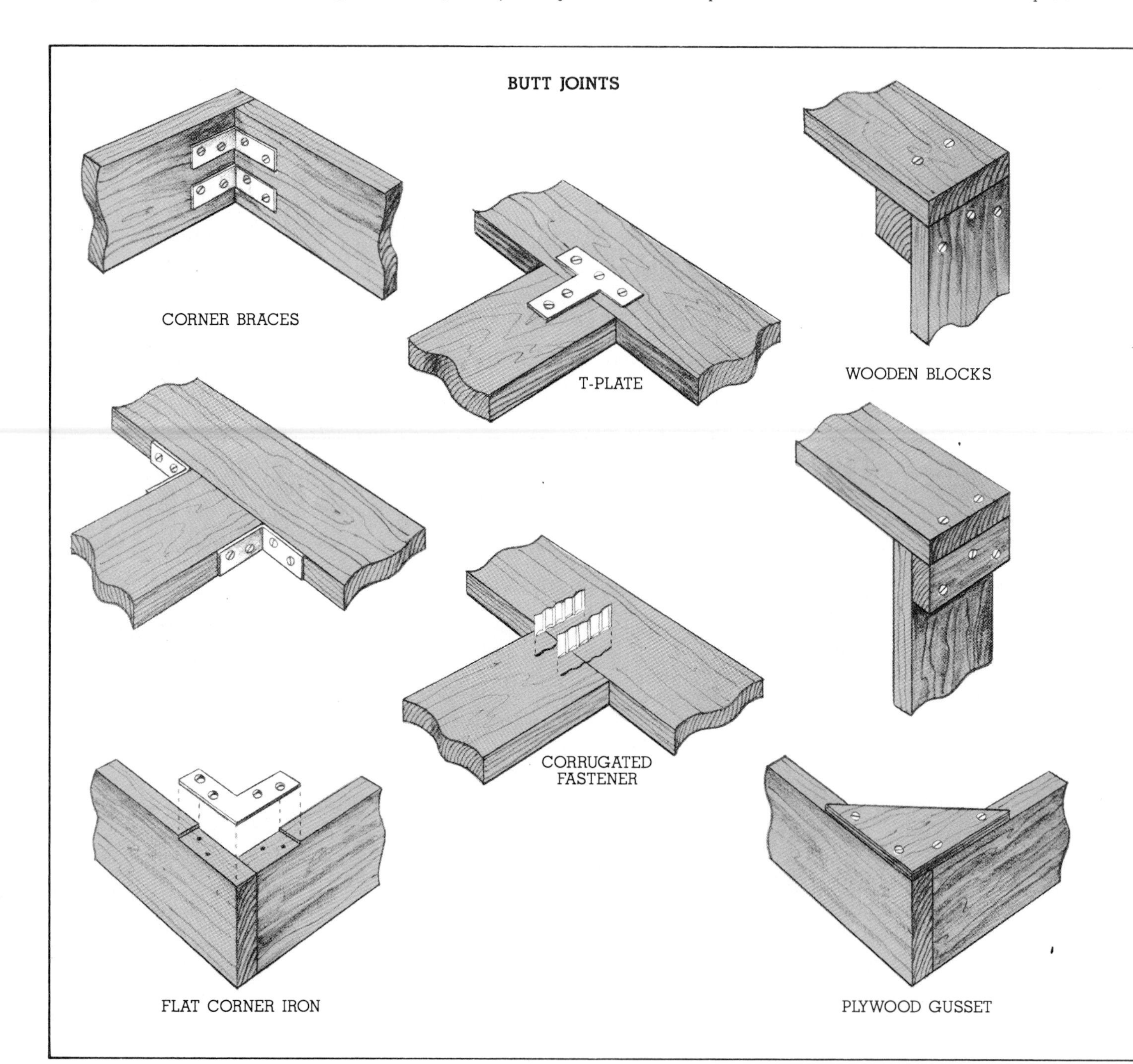

the second piece is thick. Then spread glue on the mating surfaces, position the members, and clamp until the glue dries. To fashion a *half-lap*, the strongest of the lap joints, cut a recess into each of the members that's as wide as and half as deep as each piece is thick. Glue and clamp the pieces together.

Though not quite as easy to make as butt and lap joints, *dadoes* are both attractive and strong. First, chisel a recess in one of the members—no more than one-third that member's thickness and as wide as the other member. (See pages 40-41 for chiseling techniques.) Again, apply glue and clamp until the glue sets.

For a perfectly concealed joint, a *miter* is your best bet. Begin by cutting the members to be joined at the same angle (usually 45°)—see pages 36 and 37 for how to do this—then glue and nail the members together. Note in examples 2 and 3 how you can use wooden splines and dowels to strengthen the joint. With these you eliminate the need for a metal fastener while achieving a cleaner, all-wood look. Be sure, however, to make the spline grooves and dowel holes slightly deeper than the splines and dowels in order to accommodate the glue.

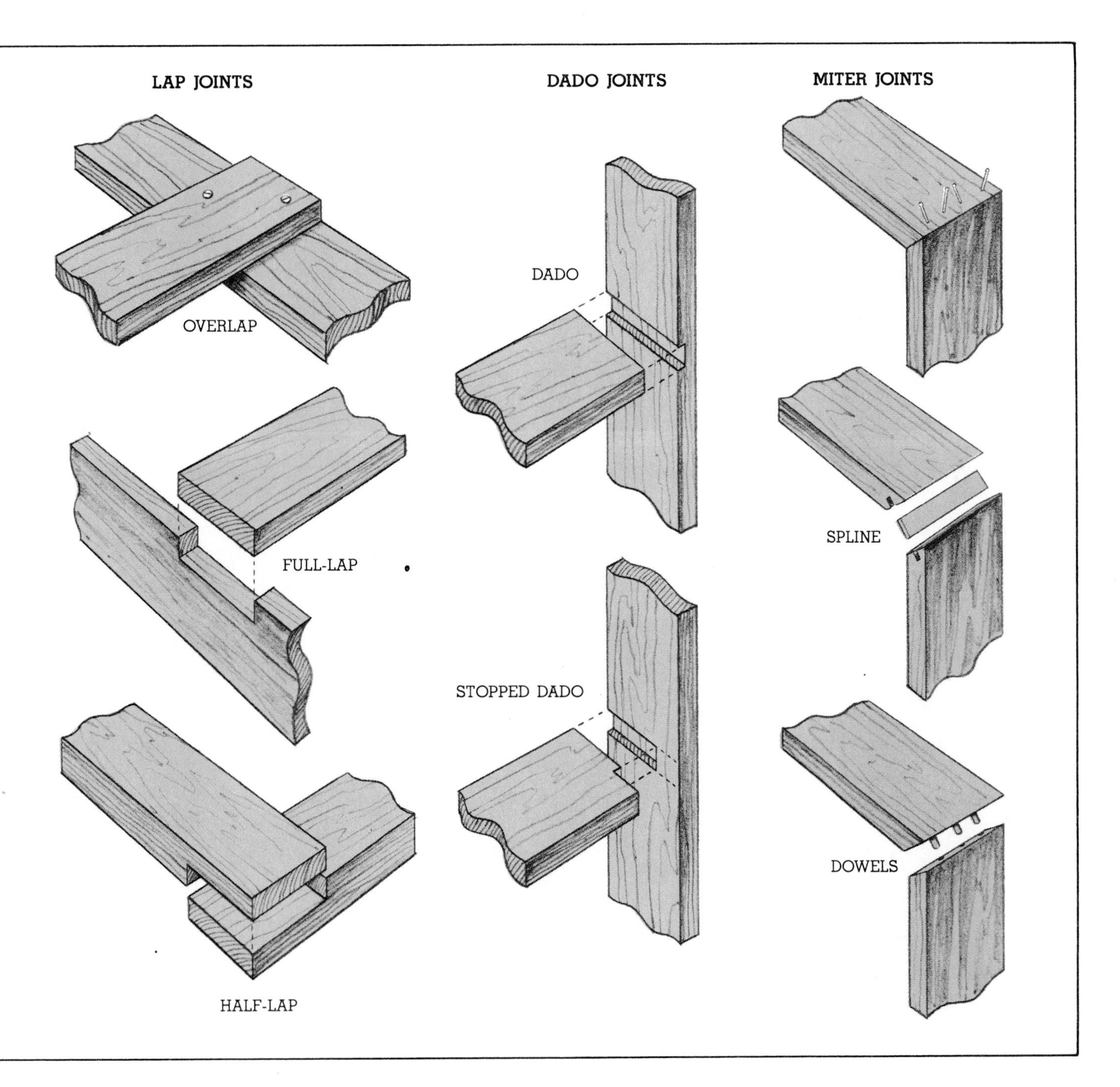

How To Hang Things on Walls

Obviously, when you hang an item on a wall, you want it to stay put. And that means you must use the correct fastener for the job. Which one you choose depends on two factors: the weight of the object to be hung, and what—if anything—is behind the wall.

1 With lightweight items, often just driving a small nail at an angle provides adequate support. For pictures and other similar wall ornaments, you'll be best off with a steel picture hanger such as the one shown here. Or if holes in walls are frowned-on where you live, a gummed hanger may be your only answer.

2 Whenever it is possible, secure medium- and heavyweight items to wall studs. To locate one of them, rap on the wall at various spots with your hand. The spaces between studs will respond with a hollow sound. A solid "thunk" indicates that you've found a stud. Or if you have walls made of drywall, you can use a magnetic stud finder to quickly locate the nailheads holding the material to the studs.

3 Tying into a single stud will give more than enough support for most items, even heavy ones. However, with large or bulky objects such as kitchen cabinets and the like, you may need to span several studs and nail a support ledger to them. To locate adjacent studs, measure over 16 inches and you should find one. (Occasionally, however, studs are placed on 24-inch centers.) To confirm this, drive a small nail through the wall material.

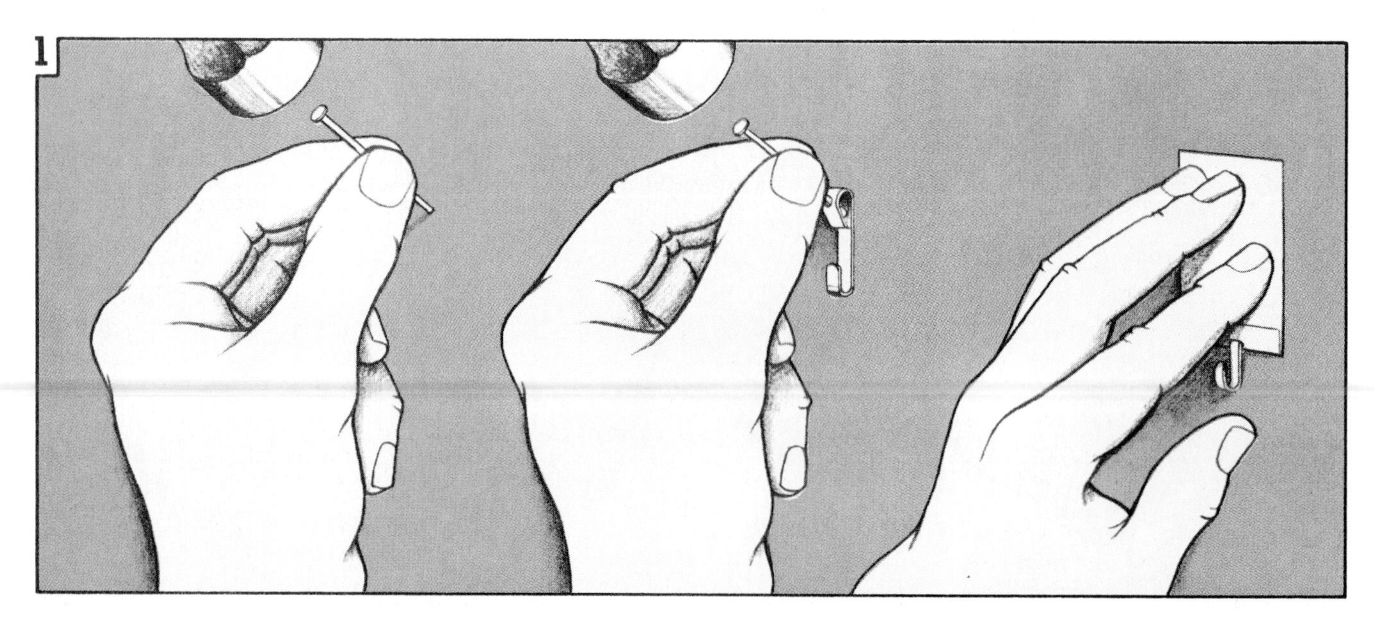

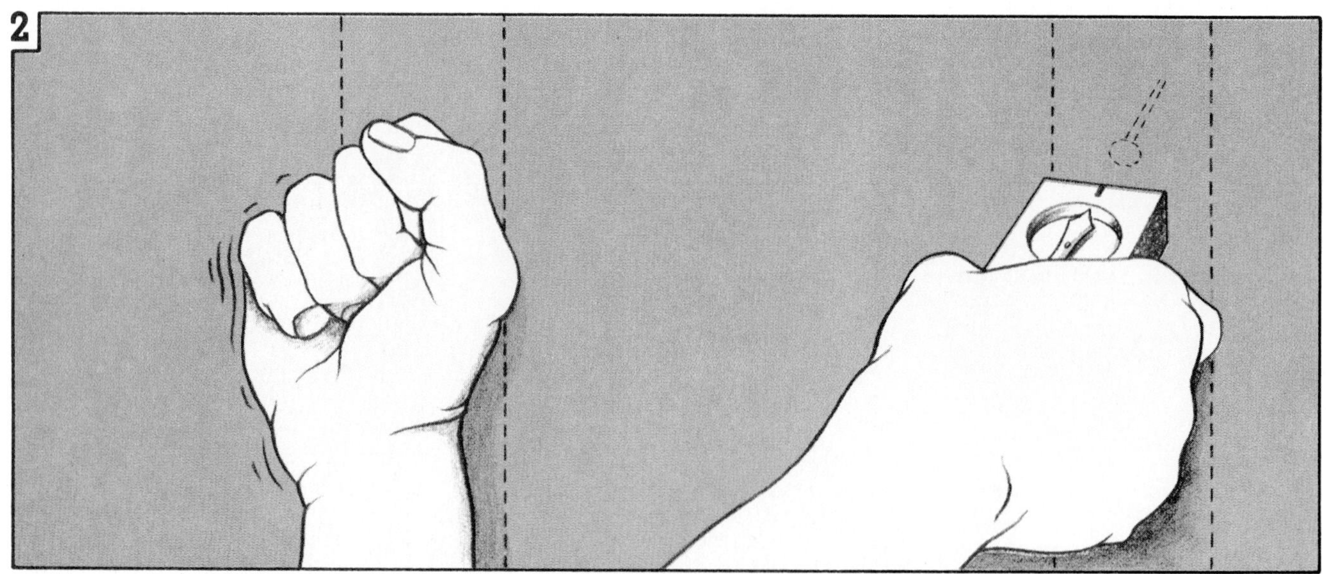

4 What if a stud isn't where you need it? Then reach for a *hollow-wall anchor* (commonly called a Molly bolt) or a *toggle bolt*. You'll find both available in a wide range of sizes; the larger the fastener, the greater its strength.

To install a hollow-wall anchor first bore a hole that's large enough to accommodate the anchor's shank through the wall material. Insert the anchor into the hole, tap its barbed flange into the wall material, and then turn the bolt clockwise till tight. As you tighten the bolt, the anchor's slotted flange collapses and grips the back side of the wall material.

Toggle bolts function in much the same way as do hollow-wall anchors, but they mount differently. First drill a hole through the wall—this one big enough to accommodate the folded-up wings of the toggle. Remove the bolt from the spring-loaded wings; slip it through a washer and then through the object to be hung. Re-attach the wings and push them all the way through the hole in the wall. As the wings move past the back side of the drywall, they'll spring out. When this happens, pull out on the fastener and at the same time, tighten the bolt.

5 Hanging things on brick, block, or stone walls isn't difficult to manage if you follow the procedures here. First, drill an appropriately sized hole into the masonry, preferably into the mortar between units, using a masonry bit. Then drive a lead, fiber, or plastic anchor into the hole with a hammer. Now, slip a lag screw through a washer, through a hole in the object you want to hang, and into the anchor. Drive the screw with a wrench. As you do so, the anchor will expand and grab the interior surface of the hole.

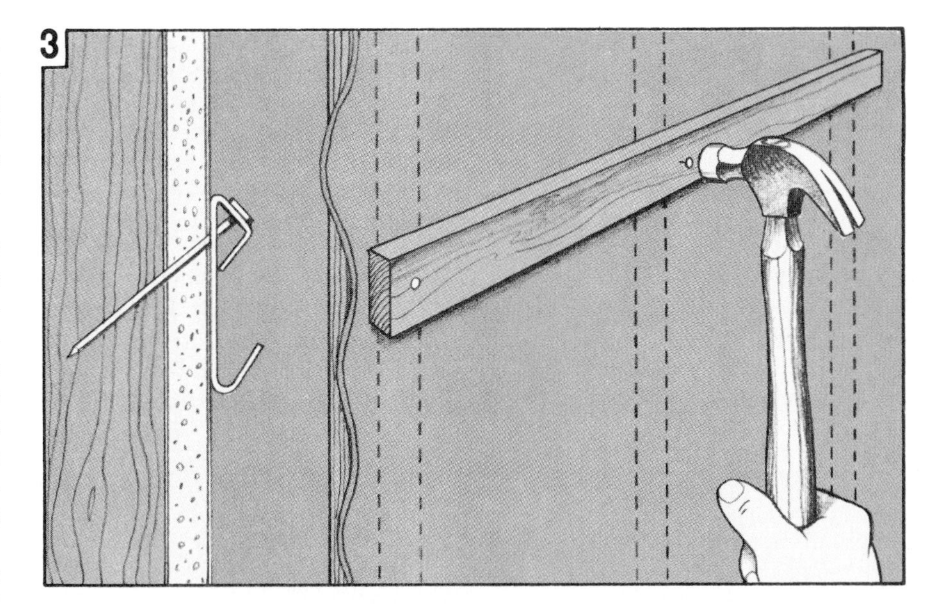

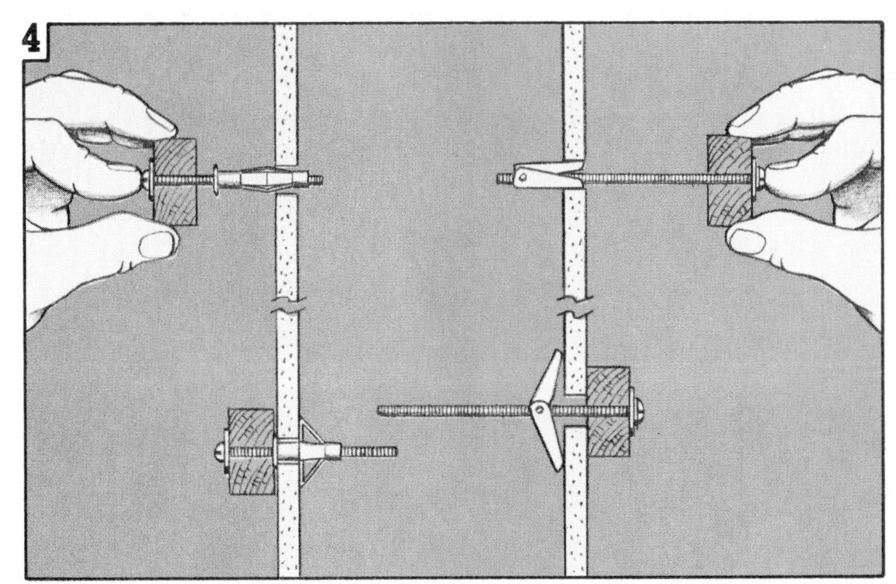

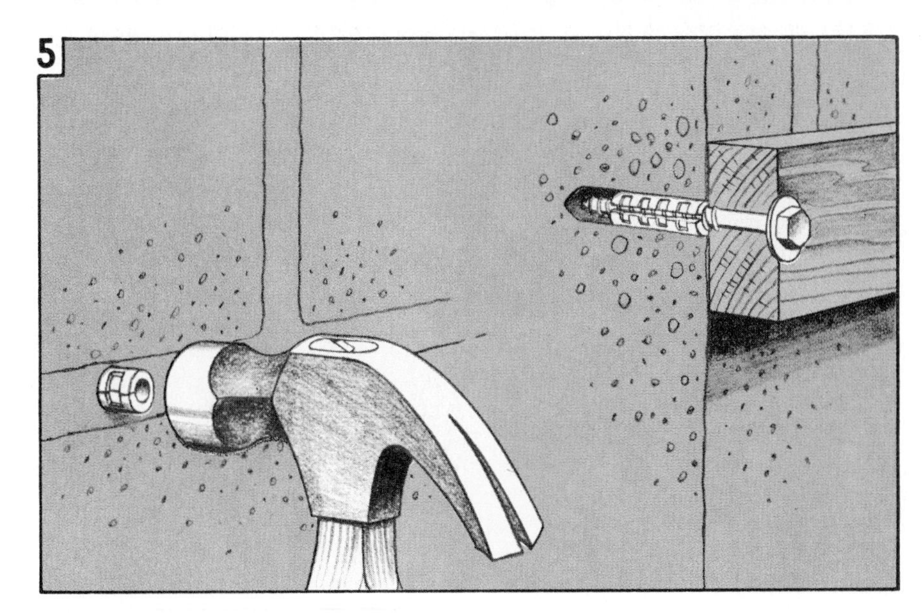

Shaping Techniques

Beveling edges and corners, planing down binding doors, trueing the edges and ends of lumber or sheet goods—you can do all these things and more once you learn how to use the various wood-shaping tools at your disposal. For most shaping jobs, three types of tools will see you through: planes, surface-forming tools, and rasps and wood files. If there's a secret to using any of them, it's keeping them clean and sharp.

However, even the sharpest shaping tools are no match for a board that's irreparably twisted, bowed, cupped, or warped (see page 14 for how to spot these defects). So, always inspect your material for flaws first, and if you find a serious one, don't waste your time and your patience trying to shape it away.

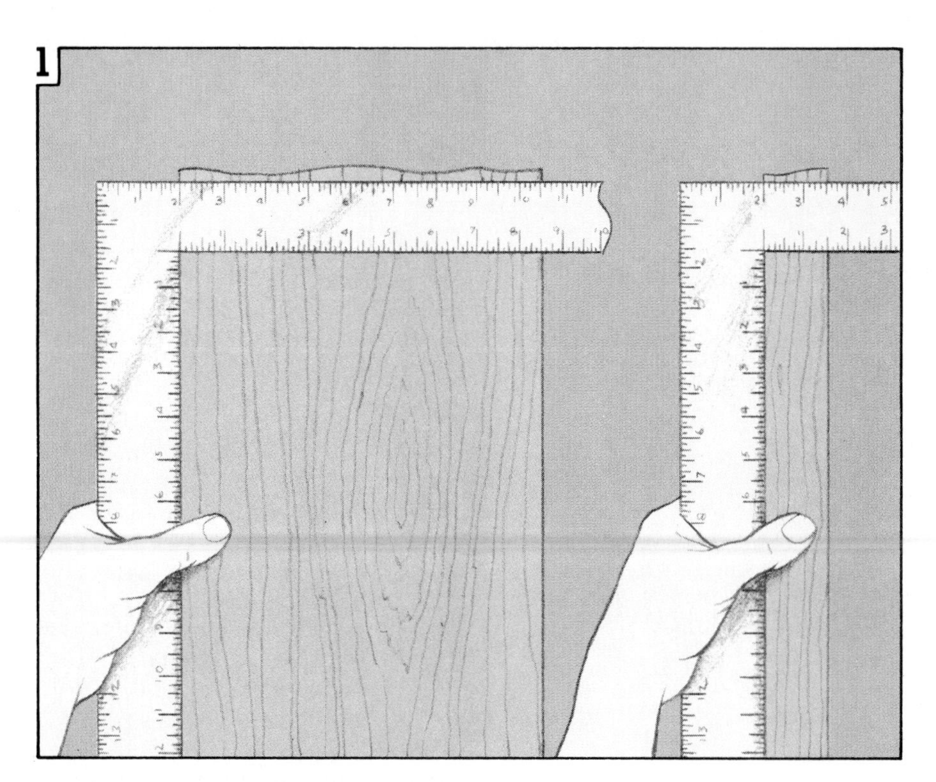

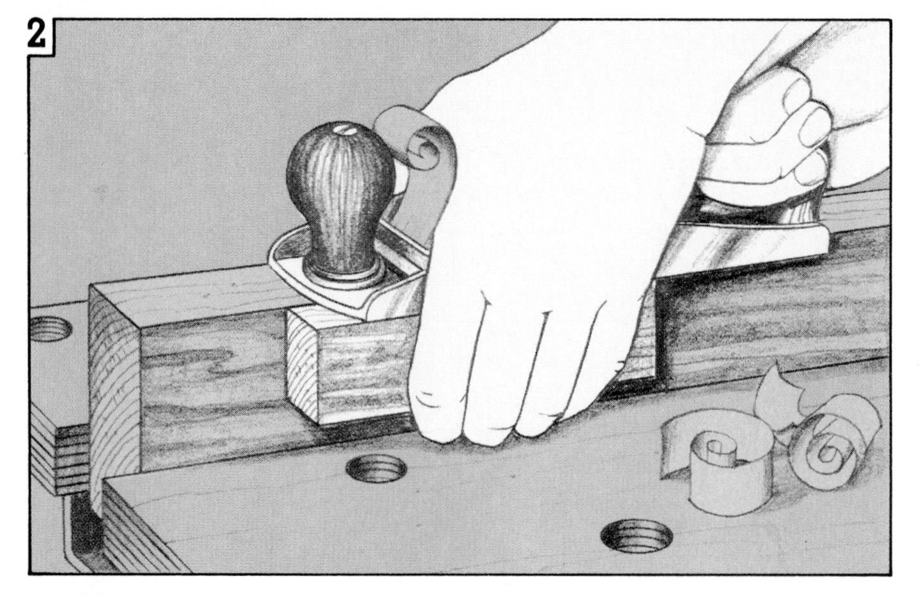

1 When smoothing almost any surface, a carpenter's framing square becomes as important as the tool you'll use to do the actual work.

To determine what needs to be done to the material to true it, first position the blade of the square along a straight, lengthwise edge and highlight the high spots with a pencil line.

Then lay the square on edge on the face of the material, and again record any unevenness with a pencil line. (You can use this same two-step procedure to shape an edge if you're sure the end of a material is true.)

2 If you need to remove lots of wood from the edge of a material, use either a jack plane or a surface-forming tool. When using either, keep these things in mind.

- First, since it takes both hands to operate the tool, clamp your work in a bench vise.
- Note the direction of the grain, and plane only in that direction.
- To avoid nicking corners, apply greatest pressure to the knob of the tool at the beginning of your cut, and to its heel at the cut's completion.
- If you plane off anything but a continuous, even-thickness shaving, a plane is either dull or adjusted for too thick of a cut, or you're planing against the grain.

Note the technique shown here for keeping a plane square to a narrow board edge. Grip a square-cornered block of wood against the bottom of the plane as shown here. Guide it firmly against the face of the board as you cut to prevent the plane from tilting on the board's edge.

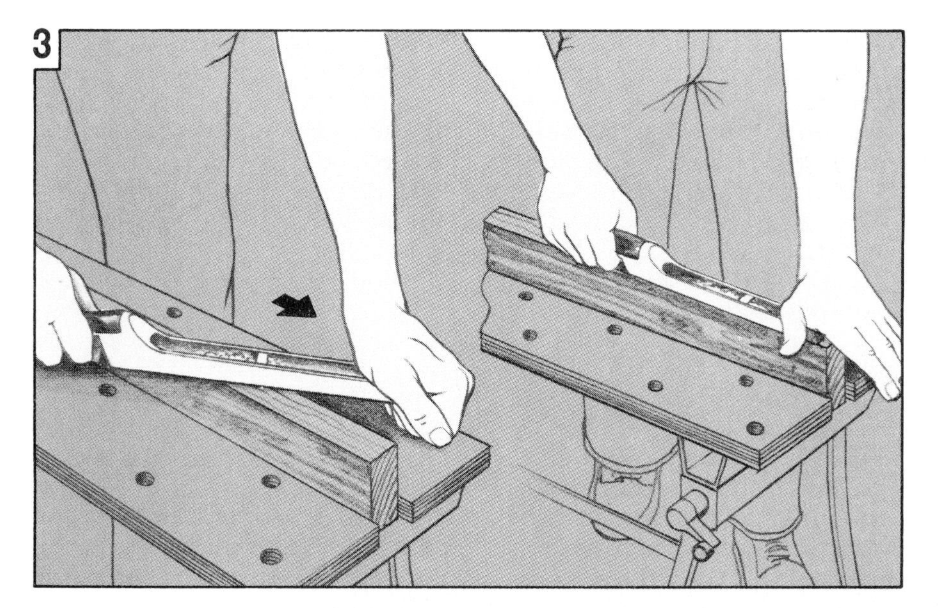

3 Surface-forming tools, also commonly known as serrated rasps, come in a variety of plane- and file-shapes and sizes. The plane-like tool shown here works much like the jack plane. How you use it differs, however. You can't adjust it for depth of cut, but you can regulate the amount of material it removes by the way you position it against the material.

For rough-cutting applications, hold the tool at a 45° angle to the work as you move it along the edge. Shavings will curl up through the serrations without clogging the holes. For a smoother result, hold the tool parallel to the board's edge.

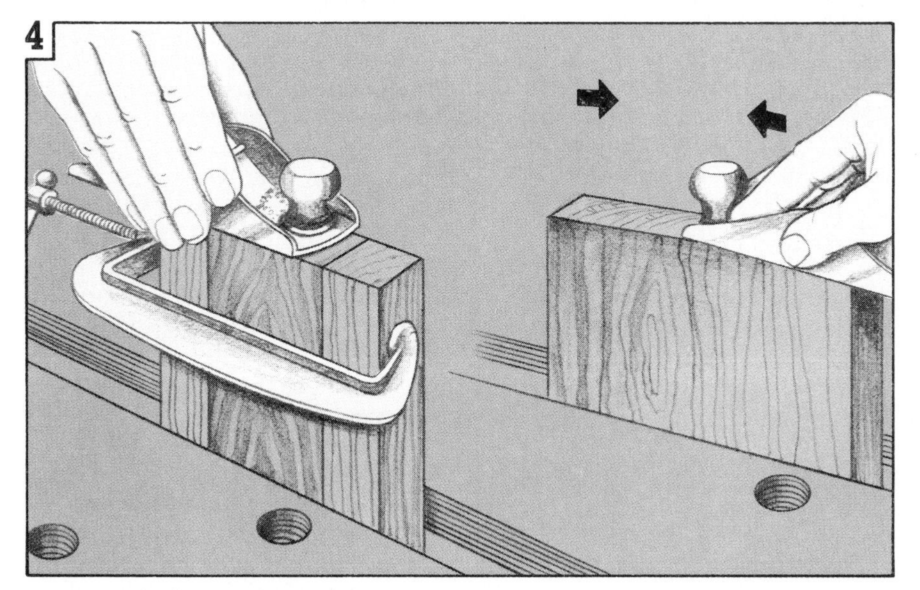

4 Shaping end grain requires both a different tool and differing techniques from those used for shaping edges. Note, too, that you must follow the procedures outlined here to keep from splintering the ends—a frequent problem whenever shaping end grain.

For narrow stock, C-clamp pieces of scrap material at both ends and plane across the end in a single direction.

Use a three-step procedure with wider stock. Begin by planing from each end only to the center. Then finish by planing off the hump in the middle, being careful not to plane all the way to the edges.

5 Wood files and rasps come into their own when you need to shape small edges or curves that planes and surface-forming tools just can't reach.

For finishing-touch shaping in which you don't need to remove a lot of material, you may operate a file one-handed, as shown here. Where you have more wood to remove, use your free hand to apply downward pressure on the toe end of the file.

Always be sure to keep your file perpendicular to the stock, and push its full length diagonally across the work, lifting it clear on the backstroke. Periodically, use a square to check your filed edge for flatness.

Smoothing Techniques

Smoothing a piece of wood is a lot like retouching a photograph: both are fine for removing minor imperfections. But if you need to go much beneath the surface—1/16 inch or more—you should do more cutting or shaping.

For best results when using sandpaper (technically called "coated abrasives") start with a "coarse" or "medium" grade, depending on the surface's condition, then finish with "fine," As mentioned earlier, aluminum oxide abrasive is the best all-around type for most carpentry needs.

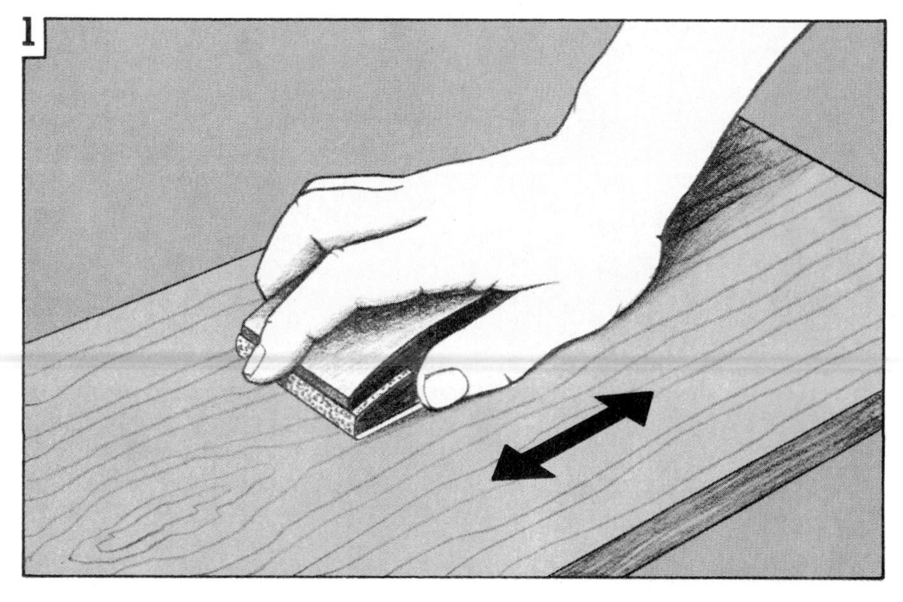

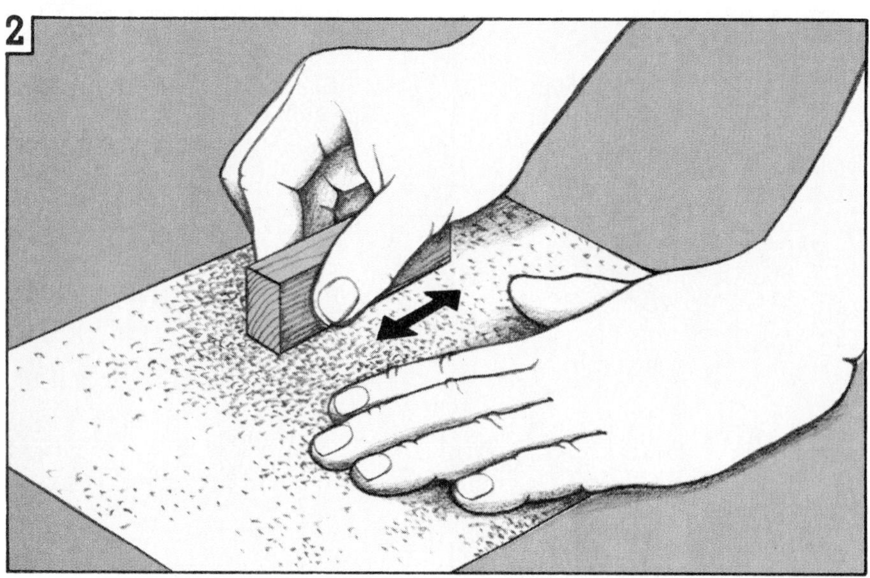

1 The simplest and best way to hand-sand a surface is with a sanding block—either the commercial variety shown here or one you can improvise by wrapping a piece of abrasive around a wooden block. Sanding with a block is less tiring than using only the palm of your hand on the paper, and it equalizes sanding pressure over a larger area for more uniform results.

When you are preparing a sanding block, tear—don't cut—abrasive sheets to size: Abrasives dull blades fast! And check that the bottom of your block is clean and smooth. Any debris trapped between the block and the paper can tear the abrasive or mar your work.

Use a sanding block only in the direction of the wood's grain. Abrasives work by tearing the surface fibers of the wood, and sanding across the grain or in a circular motion can leave hard-to-remove blemishes. If you're using the right grade of paper, light back-and-forth strokes normally are all you'll need.

2 When you need to smooth the surface of a small item, try rubbing the surface to be smoothed against a full sheet of abrasive held flat with your free hand.

Frequently inspect the surface of the work to monitor your progress, and wipe the surface with a clean cloth to prevent the abrasive from clogging quickly. When the paper finally does fill up, clean it with a few sharp raps against your workbench.

3 Because the edges of wood products are susceptible to nicks and splinters, it's a good idea to blunt them with a light sanding. Use gentle pressure and a rhythmic rocking motion. Be careful not to move the block side-to-side; the torn edge of the abrasive might catch a splinter and tear the wood.

A molded rubber sanding block like the one shown is ideal for this purpose because its base "gives" slightly under pressure. You can accomplish the same thing with a hard plastic or an improvised sanding block if you line the base with a thin layer of cork before wrapping sandpaper around it.

4 Smoothing the surfaces of round stock such as stair rails and chair spindles is both awkward and time-consuming. And if you use only fingertip pressure on a small square of sandpaper, the end result will be less than satisfactory.

To make this operation easier—and to achieve a better endproduct—use the shoeshine technique shown. Here, a cloth-backed abrasive is your best choice.

5 When smoothing wood in tight quarters, you often can get the job done with a sheet of abrasive and a little ingenuity, as in the two examples shown here.

Occasionally, you'll want to smooth two surfaces where they meet at an inside corner—neatly and quickly. A creased sheet of abrasive wrapped around a sharp-cornered block of wood is made to order.

Likewise, a strip torn from a full sheet of sandpaper and wrapped around your finger is ideal for smoothing the inside edges of circles and other small cutouts.

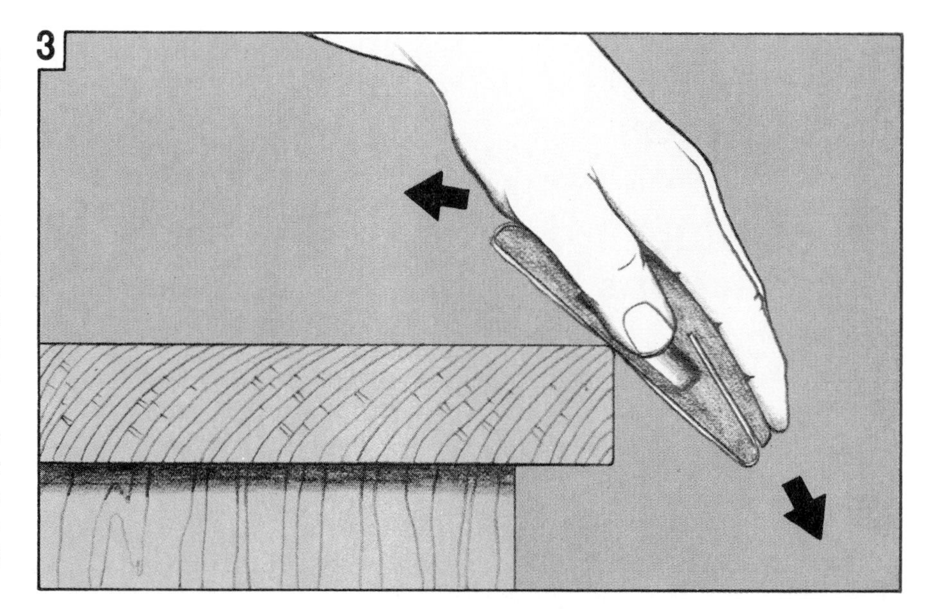

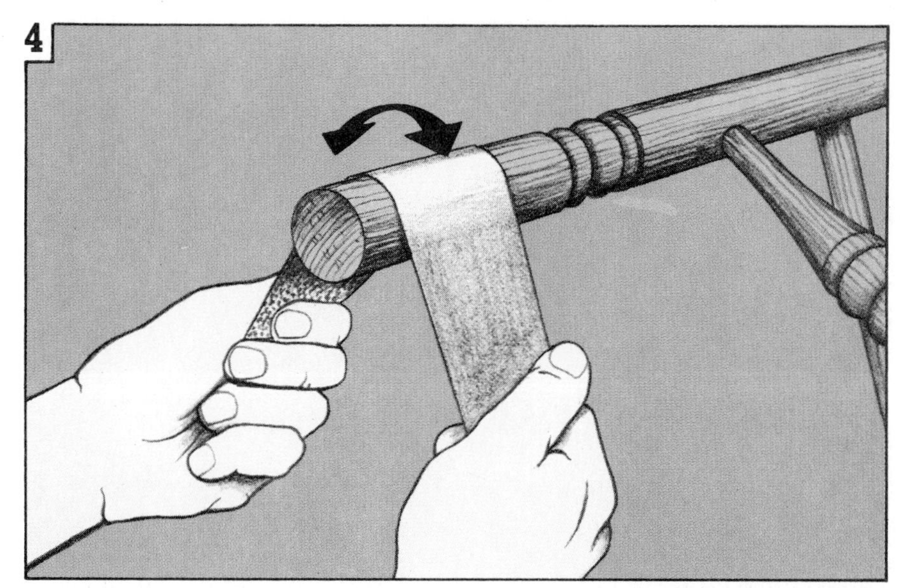

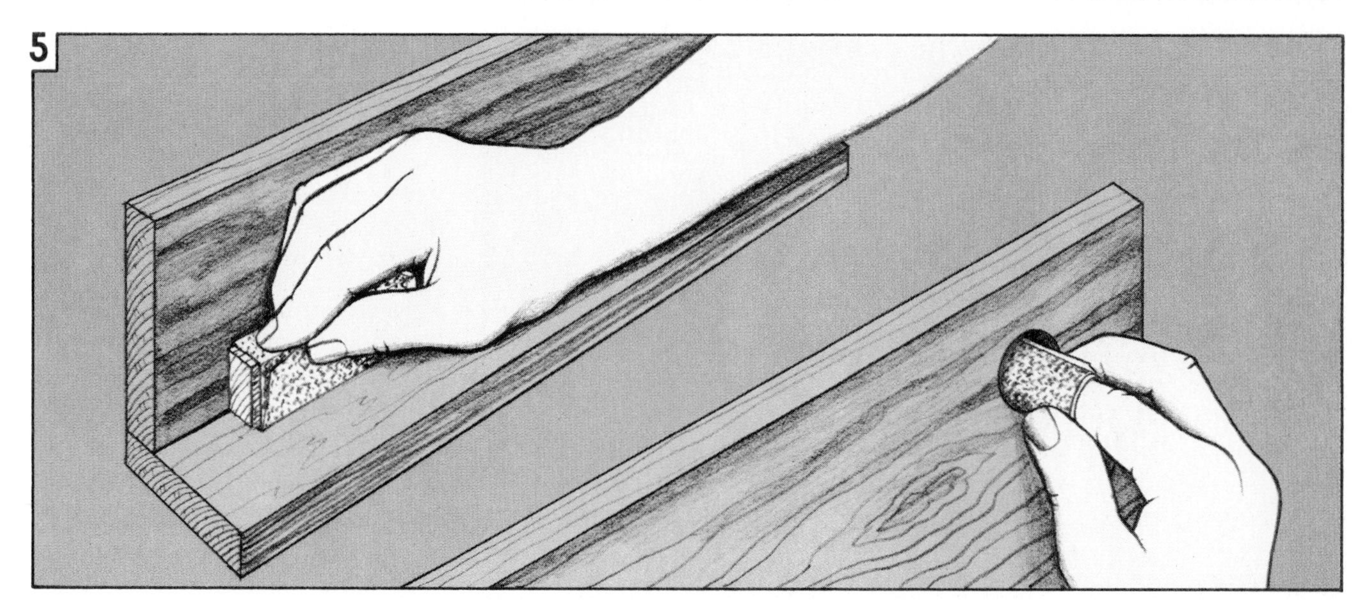

Finishing Techniques

Fill or Hide Imperfections

Often, especially near the end of a long, drawn-out carpentry project, you may be tempted to cut a few corners and slap on a finish without doing the prep work so necessary to achieve a professional-looking result. If you find yourself in this situation—and you will—remember this: No finish surface is any better than what's underneath. And in some cases, it's worse.

How you prepare your project for the finish coat depends on the type of wood you're working with and the type finish you plan to apply. If you've opted for a paint finish, simply fill any voids in the material with water putty, and after letting the putty dry, sand the surface smooth.

Preparing for a clear finish isn't quite that quick and easy. First, conceal all exposed plywood edges with wood veneer tape or screen bead molding. Then, if you're working with oak or one of the other open-grained woods, apply a *filler stain, filler sealer,* or *filler mixed with stain* to close the wood cells.

Follow this with a coat or two of sealer—shellac or "sanding sealer" works well here—to keep the stain or filler from being attacked by the varnish. Be sure to sand after each coat with very-fine grit abrasive.

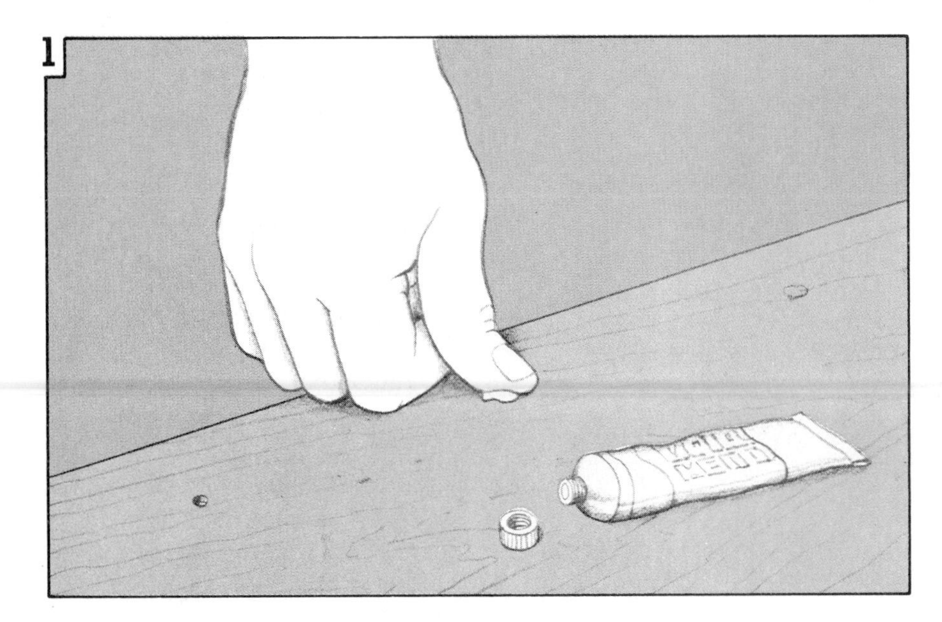

1 To fill voids in wood you plan to paint, such as nail holes and counterbored screw holes, a plastic dough-type filler is a good choice. Simply tamp a small amount of it into the hole with your thumb. As the solvent in the dough evaporates, the filler will harden and become sandable.

For very deep or large voids, apply two or three layers of filler, letting each dry completely before filling further. Let your final layer overfill the void slightly to allow for the putty's tendency to shrink as it dries, and to leave a base for sanding.

2 Water-mix putties excel at filling shallow depressions over a large surface area. (Again, use this type of filler only on projects you'll be painting.) By mixing in a little more water (or a little more powder), you can achieve just the right spreading consistency. One caution, however: Water-mix putty sets up quickly, so don't mix more than you'll be able to use in 10 or 15 minutes.

To blend in a knot with its surrounding wood, mix the putty to a pastelike consistency and apply it with a putty knife. Force the mixture into all cracks, feather the edges of the patch to the surrounding wood, and sand smooth when dry.

To fill the voids on plywood edges (or to level the end grain of

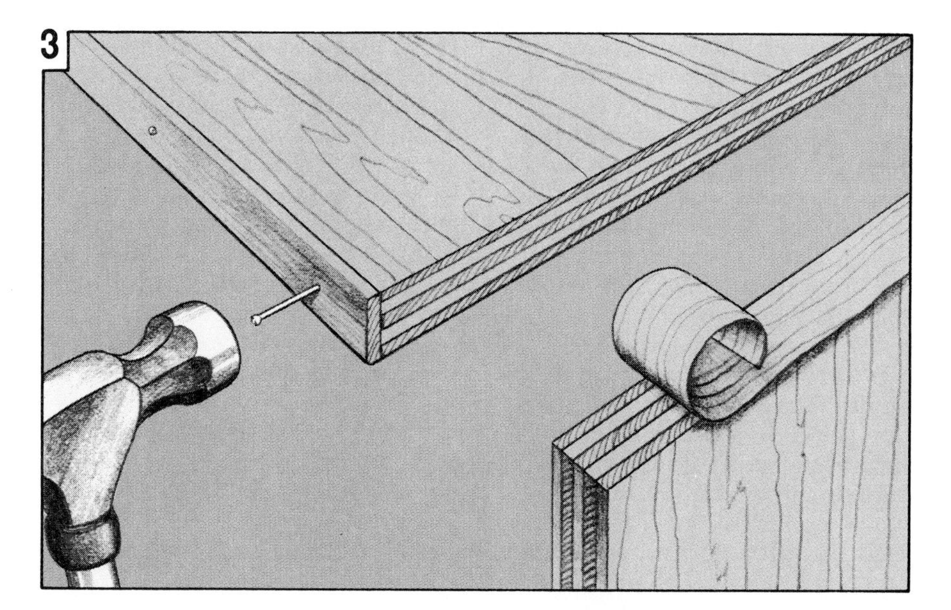

boards), mix the putty to a slightly thinner consistency. When it's completely dry, sand with a sanding block. For a really smooth surface, apply a thin second layer of filler.

3 If you plan to clear-finish the project, you'll want to finish the edges of plywood in either of the ways shown here. To conceal an edge with *wood veneer tape*, simply cut a length of it with scissors and apply it to the plywood's edges with contact adhesive (some self-adhesive types are applied with a hot iron). Purchase tape that is wider than the material is thick and that matches the veneer of the plywood. Trim the excess tape from the edge with a sharp chisel or pocket knife.

You also can hide unsightly plywood edges with inexpensive *screen bead molding*. To do this, butt the molding to the plywood edges (preferably using glue), and nail it in place with finishing brads. Sink nailheads below the surface. Don't fill the voids yet; you'll do this with color-matched wood putty after staining the project.

4 When you want a glassy smooth finish on fir, oak, or other open-grained woods, don't waste your time trying to sand the surface smooth. Instead, tame pronounced wood grain with *paste wood filler*.

Use a clean, dry brush to work the filler into the surface from all directions. After 10 to 15 minutes, level the filler by dragging a piece of cardboard across the wood grain. Finally, when the filler is almost dry, carefully drag the surface again—this time with the wood grain.

5 After filling or staining, apply a coat or two of shellac thinned with denatured alcohol, or sanding sealer.

Using a clean, dry brush, work it into the wood from several directions, but always finish brushing with the grain, using long, even strokes.

Apply the Finish Coat

A correctly applied paint or clear finish does two important things: It beautifies and protects. And providing you've prepared the wood surface with care, this last step in the finishing process can be one of the most satisfying parts of your project.

Unless you've been making a furniture-grade item, you'll probably go with a paint finish. In this case, first apply a water- or solvent-based primer to the bare wood—one that's compatible with the topcoat you plan to apply. Primers ready the wood to accept the topcoat and serve to fill the wood pores. Allow the primer to dry, sand the surface with very-fine abrasive, and lay on one or more coats of paint.

For showy projects that have a beautifully grained wood you want to highlight, choose one of the many clear finishes available as your protector/beautifier. If you want to darken or lighten the wood's color, stain or bleach it first, then seal the surface with a sealer. With open-grained woods, you may want to apply a filler stain (see page 62), then a sealer. Whatever method you use, allow each coat to dry thoroughly and sand smooth with very-fine abrasive. Complete the project by applying two or more coats of varnish or hand-rubbed oil.

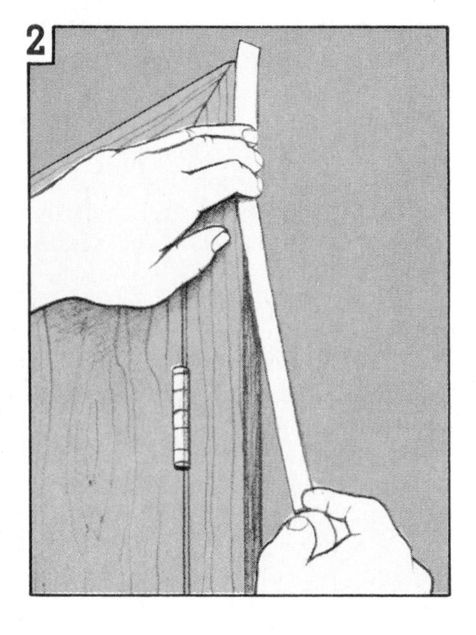

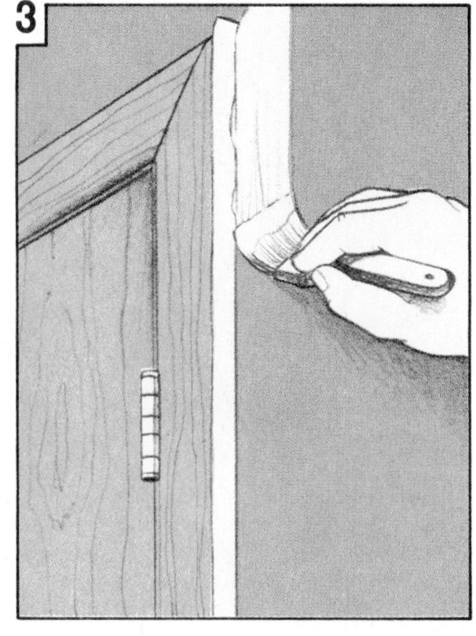

1 Painting with a brush is a familiar task, but perhaps not so familiar that a few work-saving tips aren't in order.

Begin applying paint to wood surfaces with short strokes across the wood grain, laying down paint in both directions. Don't bear down too hard on the bristles. Finish painting with longer, sweeping strokes in one direction only—this time with the wood grain. Just use the tips of the bristles to level the paint.

This two-step technique ensures complete coverage of the surface. This technique is useful for applying varnishes, too.

2 Professional painters have had years of practice to perfect the technique of freehand painting; they can guide a sash brush along razor-thin lines with hardly a second thought. For the rest of us, however, some judiciously placed masking tape can bring about equally satisfying results.

To cut-in wall surfaces around woodwork, use tape to conceal all adjacent wood surfaces, butting it against the wall.

3 Now work your brush all the way into corners (you needn't worry about spilling over onto the wood).

Masking windows and woodwork can save you hours of tedious cleanup. Just be sure to remove the tape while the paint is still tacky; otherwise, you risk leaving a ragged edge where the tape meets the walls, as well as difficult-to-remove tape.

4 For painting large areas, a roller is the tool to use—and this is how to use it to achieve complete, uniform coverage. Load the roller with paint and begin applying it in a large "M" shape. Start with an upward stroke, increasing pressure to squeeze out more paint as you go. Cross-roll with horizontal strokes to level the paint and to fill in between your diagonal strokes.

5 Spray-painting with an aerosol can is fast and easy, but potentially messy. Minimize the mess by masking off adjacent areas with newspaper. Or, when spraying small objects, improvise a painting booth from cardboard and clear plastic. To avoid breathing airborne paint, always wear a painter's mask.

6 Stains can beautify almost any wood, and are among the easiest finishing products to use—just brush them on, then wipe them off.

Use a clean, dry brush to apply stain in the direction of the wood grain. Let it stand for a few minutes, then simply wipe it off with a clean rag. To achieve a darker hue, let the stain stand longer before wiping, or apply a second coat. If you've already gone too dark, you can rub off some of the pigment with a cloth moistened in the thinner recommended for the stain you're using.

7 Any wood finish you apply in successive coats will look better and last longer if you lightly abrade the surface between coats. Whether you're between coats of paint, coats of varnish, or a sealer coat and a lacquer, a light smoothing with very-fine abrasive gives the next coat an acceptable surface to cling to and makes it easier to see where you've been as you apply the next coat.

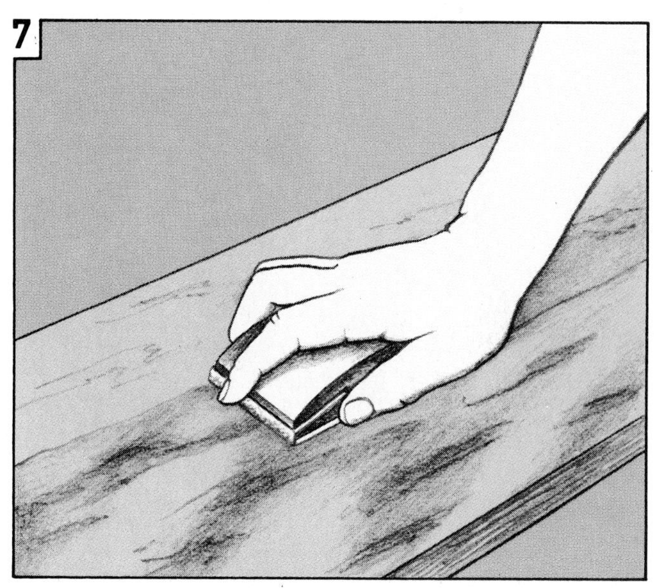

Tool-Sharpening Techniques

A do-it-yourselfer of our acquaintance once observed that working with a dull tool is a lot like trying to roller skate uphill with the wind in your face. How true! But if you heed the sharpening advice on these pages, you needn't fight uphill battles against balky tools.

When a tool dulls, your sharpening goal is to remove any nicks and burrs and to restore the cutting edge to its original factory angle. Usually, you can do this by giving it a touch-up.

But in the case of saw blades, after several touch-ups, the teeth will need *setting*—rebending the blade's teeth to the proper angle. This kind of work—and the sharpening of circular saw blades—is best left to a sharpening service.

As easy as it is to restore most cutting edges, sometimes it's better to replace a worn-out tool than to resharpen it. Unless you have top-quality drill bits, for example, you may want to buy a new bit rather than take time to sharpen the old one.

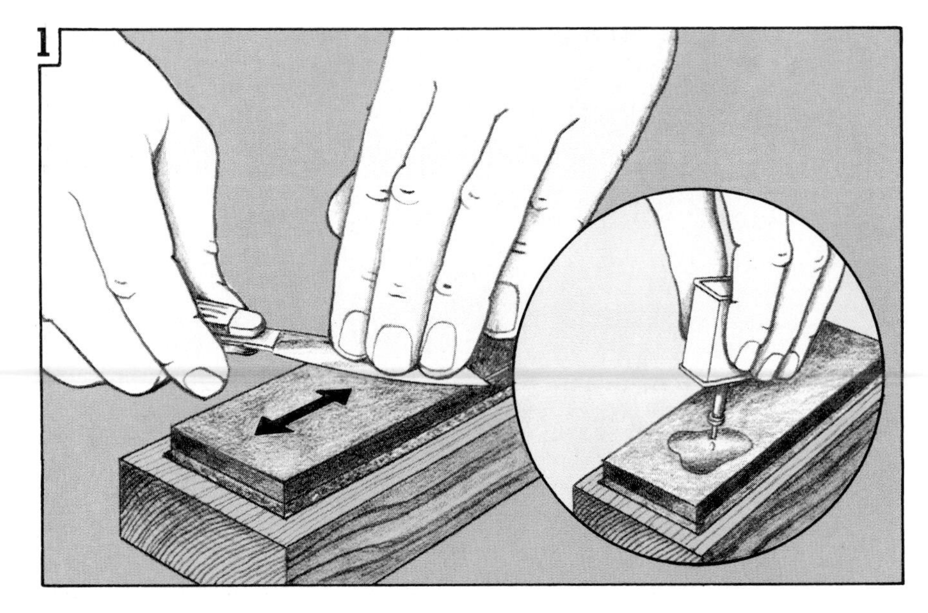

1 Sharpen the blades of pocket- and small utility knives (but never sawtooth or serrated blades) with a two-grit whetstone. Hold the blade at a 30-degree angle to the surface of the stone, and in such a way that the blade travels over the stone at a slight diagonal. Wet the stone with a few drops of honing oil, then stroke the blade across the surface, rocking it from heel to blade tip. Turn blade over and stroke in the opposite direction the same number of strokes.

2 Dull plane irons respond well to touch-up sharpening with a whetstone, too. Remove the iron from the plane and use the following two-step procedure.

First hone the beveled edge of the plane iron with back-and-forth strokes on the whetstone. Try to maintain the factory-set angle—approximately 30 degrees. Now turn the iron over, lay it flat on the whetstone, and move it in a circular motion to remove burrs that may have formed on the flat side.

3 You hone chisels in exactly the same way as plane irons (see opposite page). But when their tips become badly nicked or blunted, file them first.

Lock the chisel in a bench vise as shown, and file off the worn cutting edge with a file held perpendicular to the chisel.

4 Now restore the cutting edge by holding the file at the angle of the chisel's bevel and slowly filing diagonally across the bevel.

5 When you seem to be sawing harder and cutting less with a handsaw, it's time for some touch-up sharpening. Clamp the saw in a bench vise so the blade's teeth are about ⅛ inch above the scraps, and run a flat file across the points of the teeth until you've made small "plateaus" on the tips. Now select a small triangular file and use one of the following two procedures.

For crosscut saws, begin at the tip and place the triangular file in the *gullet* (valley) to the left of the first tooth that's set toward you. Seat the file against the bevel of the tooth (it will be at a 60-degree angle to the saw) and file until you've removed half of the neighboring plateaus you made earlier. Hold the file at both ends, and keep it perfectly level. Now skip the next gullet, file again, and continue in this manner.

When you reach the handle, reverse the saw in the vise and resume filing in the first gullet to the right of the tip-most tooth set toward you (the first gullet you skipped on the other side). File away the remaining plateau halves on each side of the gullet, and file alternate gullets until you reach the handle.

Sharpen ripsaws in the same manner, except keep the file perpendicular to the saw blade. File the teeth set toward you, reverse the saw, and complete filing.

6 Sharpening a twist drill bit requires a bench grinder with a tool rest to help you maintain the precise 59-degree angle with the grinding wheel. Holding the bit at the proper angle, slowly guide the leading edge of one of the cutting lips into the wheel. Upon contact, rotate the bit clockwise, slightly raising the chuck end of the bit and moving it to the left. This tricky maneuver grinds the heel of the bit's lip at a slightly greater angle so only the lip's leading edge cuts. Repeat this procedure with the opposite cutting lip.

Sharpening a spade bit is easier. Just touch up the shoulders with a flat file, holding it at the angle of the bevel—about 8 degrees.

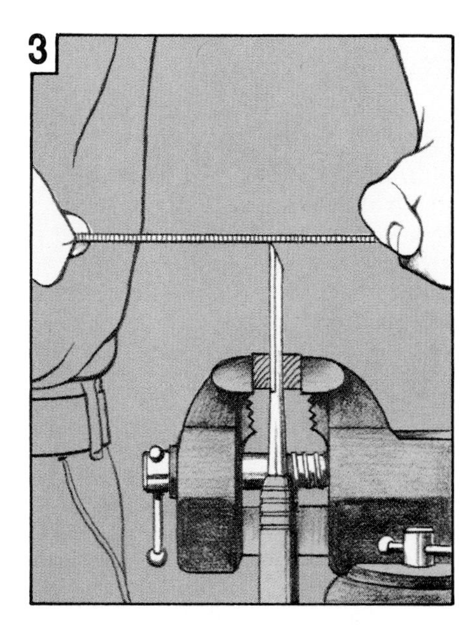

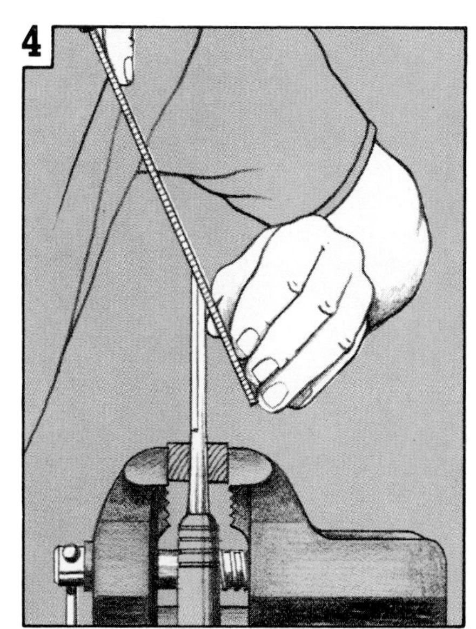

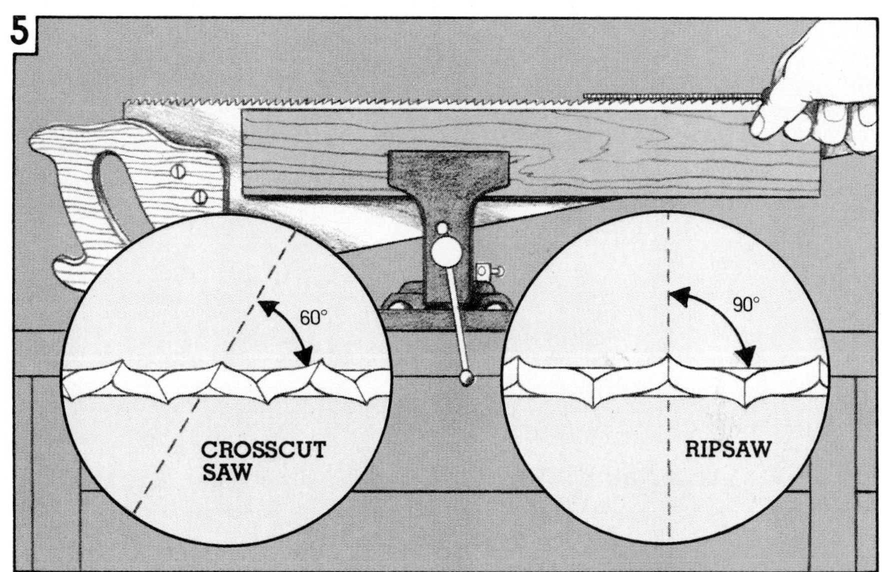

TRY THESE BASIC PROJECTS

What is knowledge without some plans to put that know-how to work, right? That's why we've included here a sprinkling of projects designed to prove to you that you CAN do some substantial carpentry work and do it quite well.

Because so many carpentry projects require a knowledge of box building, that's where we start. We also show you several clever ways to hang shelves on walls, as well as how to frame walls, install furring strips, hang drywall, and install paneling.

But don't confine your thinking only to those projects we've included. With the knowledge you have garnered from the previous two chapters, you can handle a great many of the carpentry creations you'll run across in the various shelter magazines, see along the streets, or otherwise be exposed to.

Simply take the time to plan the project carefully and discuss the undertaking with your building materials supplier, if necessary. Then, providing you take an exacting approach to the work, you can't miss.

How To Build Boxes

Experienced carpenters and home builders know it. So do designers and architects. But you may not realize the significance of the basic box. It's *the original* basic building block, and once you know how to construct one, you'll be amazed at how many things you can make.

Just take a look around your home, and you can't help but be convinced of this. Cabinets, cupboards, bookshelves, and chests, to name just a few common examples, are simply boxes with a few extras thrown in.

For most of your box-building projects, you'll find plywood or particleboard the best material to use. With either of these, you're not restricted to a certain size box, as is the case when building one with solid wood. In addition, their large size and factory-straight edges mean fewer cuts for you to make.

To avoid errors at the expensive construction stage, sketch your box, including dimensions, on a piece of paper as a guide. Plan your saw cuts so that the grain of the pieces runs the same way after assembly. Take care, too, that all edges after sawing are straight and that corners are square. This prevents daylight from peeking through the joints, and keeps the box from wobbling.

As you might suspect, boxes go together in a variety of ways. Equilateral butt-jointed boxes feature quick construction; their cousins, inside butt-jointed boxes, give strength and durability. Miter-jointed ones, on the other hand, excel at hiding unattractive edges. Keep these features in mind when designing your box.

For information on strengthening butt and miter joints, see pages 54-55.

1 Note the butt joinery employed here. The sides share identical dimensions and fit together edge to side, side to edge. Measuring and sawing for this construction are a snap.

2 In our second basic box, two sides are shorter by the thickness of the other two. For instance, if you use ¾-inch plywood for a box of this type, cut two sides 1½ inches shorter than the other two.

The joinery here—edge to side, edge to side—makes for a sturdier construction than our first box, since the inside pieces are locked into place by a row of fasteners at each joint.

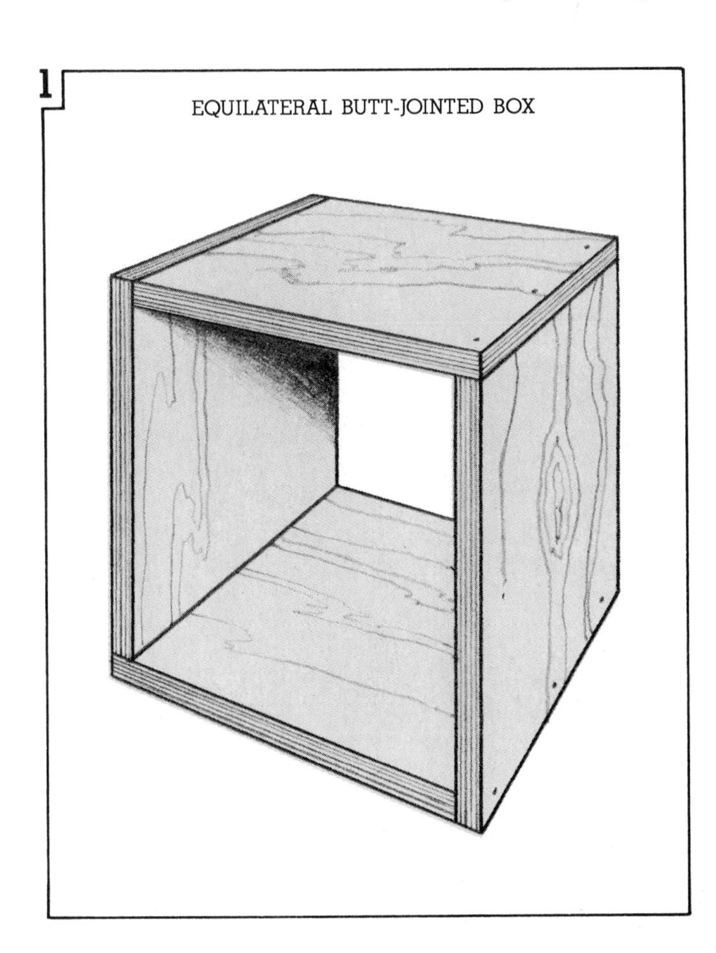

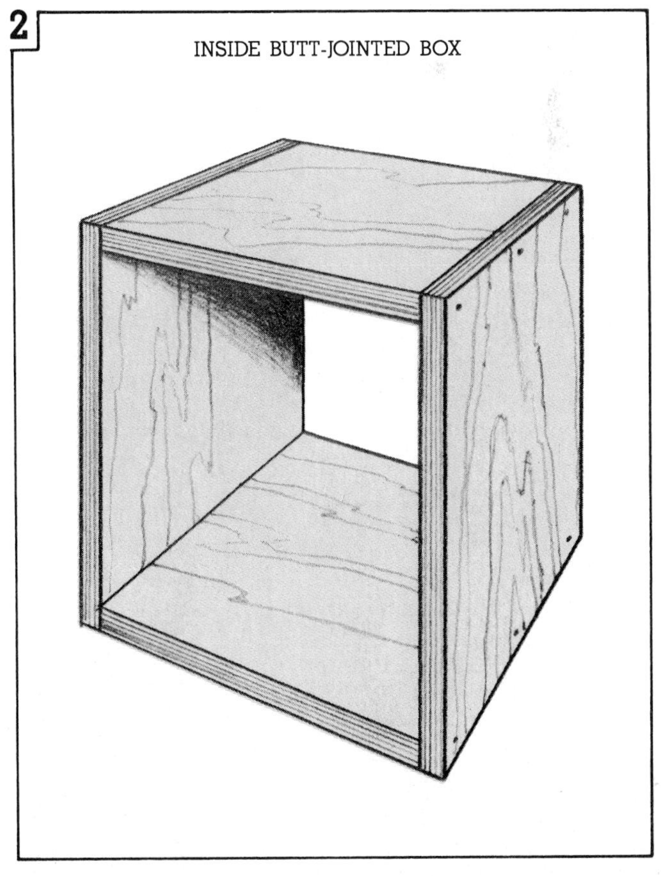

How To Build Boxes *(continued)*

3 After you've cut the pieces of the box and selected a fastener, apply glue and begin assembling. Ensure that the pieces are flush by feeling for any unevenness along the joined edges with your fingers. Then square the corners with a square.

Now drive in your nails or screws, beginning at the corners and then filling in between. If using screws, drill pilot holes first. Go back and set the nails with a nail set.

If you're having trouble joining the first two pieces, lay one member flat and against something solid, such as a wall (see the detail). Butt the adjoining side perpendicular to it and proceed as already described.

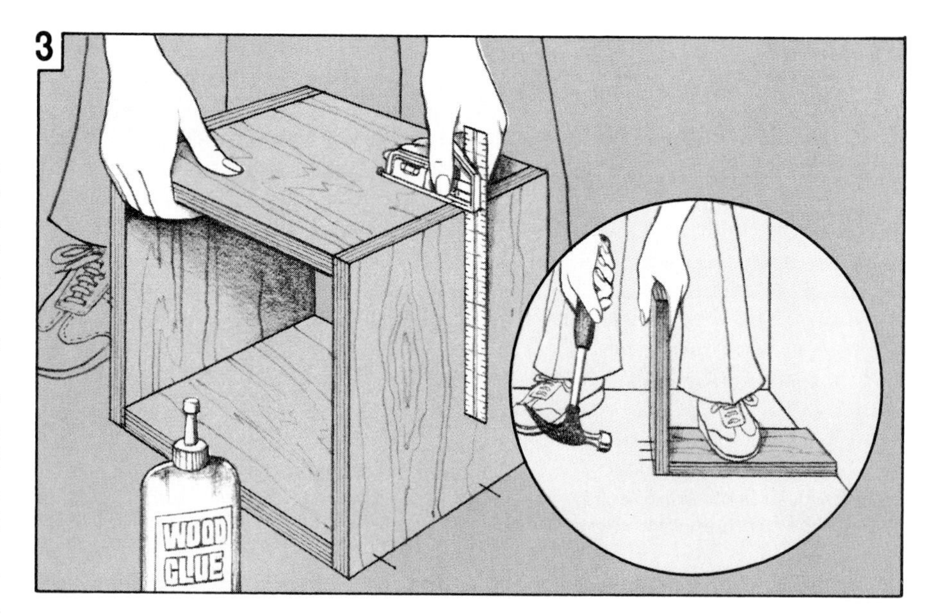

4 Boxes with mitered corners provide still another way to go. Though not as strong as the standard butt-jointed box, they avoid the problem of exposed board ends and are therefore more desirable for finish work. Bevels may be cut with a back saw and miter box for shallow boxes, and with a circular saw or saber saw adjusted to the appropiate angle for wider material. (For information on making bevel cuts, see pages 36-37.)

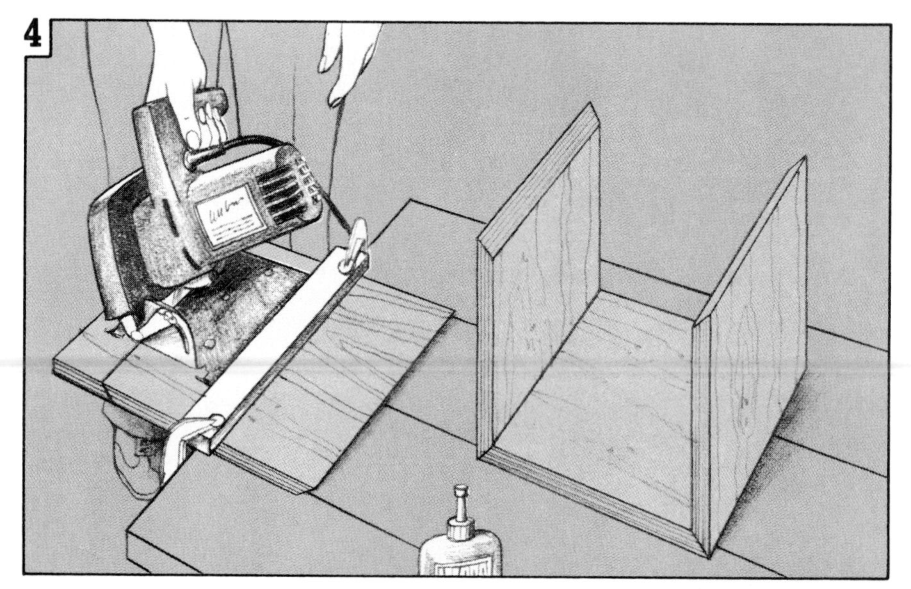

5 If you elect to add a back to a box, you have two simple options. Choose either a piece that fits neatly against the inside perimeter of the box (an *inset back*) or one that's as wide as the box's outside perimeter and fastens flush to the edges. Adding a back strengthens the sides and locks the corners together.

To measure for a back, simply place your box on an adequately sized piece of material, square the corners, and pencil a cut line along the box's inside or outside edge as desired. (If you later plan to hang your box on a wall, turn to step 5 on page 73.)

Adding Shelves to Boxes

Now that you're an accomplished box builder, consider adding some versatility to your creation with shelves. Look over the well-rounded selection of shelf support systems below and pick out one that best fits your tastes and storage needs. Also decide on the number of shelves you need for your situation and whether you want them to be adjustable or fixed.

The popular *standards* and *clips* used for adjustable shelves nail or screw vertically to the sides of boxes (for a cleaner look, recess the standards). When inserting the metal clips, do so with taped pliers to protect their surface.

Sturdy, adjustable *pin-type clips* fit snugly into pre-drilled holes ⅜ to ½ inch deep (see page 43 for instructions on drilling holes to a specific depth).

The third and least expensive adjustable shelf support system, *dowel pegs* also go into pre-drilled holes ½ inch deep. With these, the thicker the pegs, the more support.

Shelves screwed in place through the box's sides head the list of non-adjustable types. For a finished look here, countersink or counterbore the screws, then fill them with putty, wood plugs, or buttons as desired.

Dadoed shelves, though time-consuming to make, offer a sleek, functional alternative. Make the horizontal grooves in the sides *before* assembling the box.

For the strongest shelves of all, go with 1× *wood cleats* glued and nailed or screwed at the desired height.

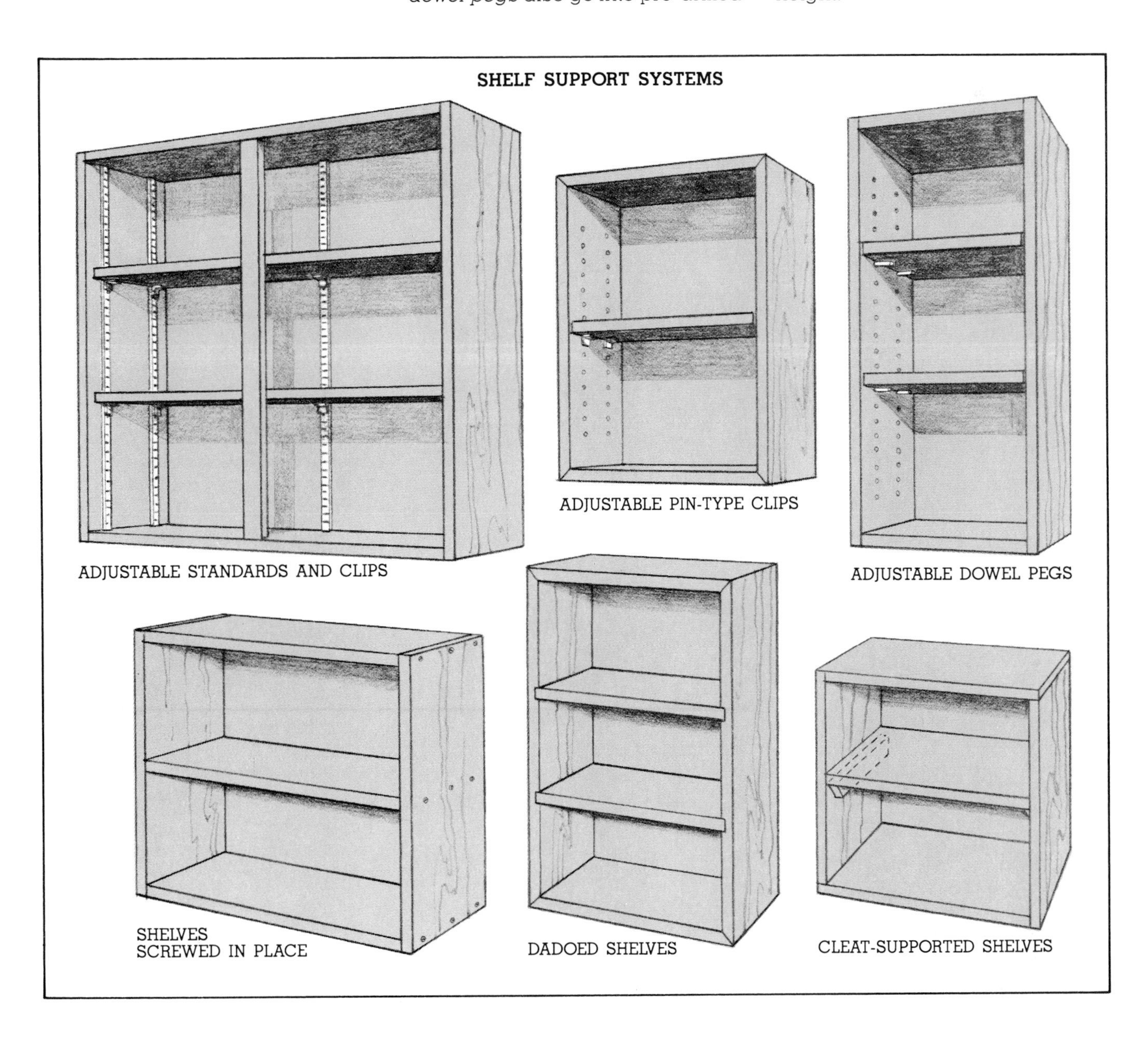

Hanging Shelves on Walls

Though definitely not one of the more glamorous carpentry projects you'll ever undertake, hanging shelves may well be the most functional. And fortunately, it's among the easiest as well.

Naturally, one of your first considerations should be how best to support the shelves. See page 24 for some of your shelf support options. Also see pages 56-57 to find out about the appropriate fastener for your situation.

Next, to prevent the shelves from sagging, you need to determine the correct distance between shelf supports. For help with this, refer to the span chart here. The spacings listed assume shelves fully loaded with books—the heaviest objects you're likely to put on them.

And lastly, to figure the vertical distance between shelves, measure the largest items slated to go on the shelves, and add at least an inch for overhead clearance.

On this and the following page we show you how to mount shelf standards and cleat-mounted shelving units. But much of the information presented applies to the other shelf support systems as well.

Shelving Spans

Material Used	Maximum Span	Material Used	Maximum Span
¾-inch plywood	36″	2×10 or 2×12 lumber	48-56″
¾-inch particleboard	28″	½-inch acrylic	22″
1×12 lumber	24″	⅜-inch glass	18″

(Assumes shelves fully loaded with books)

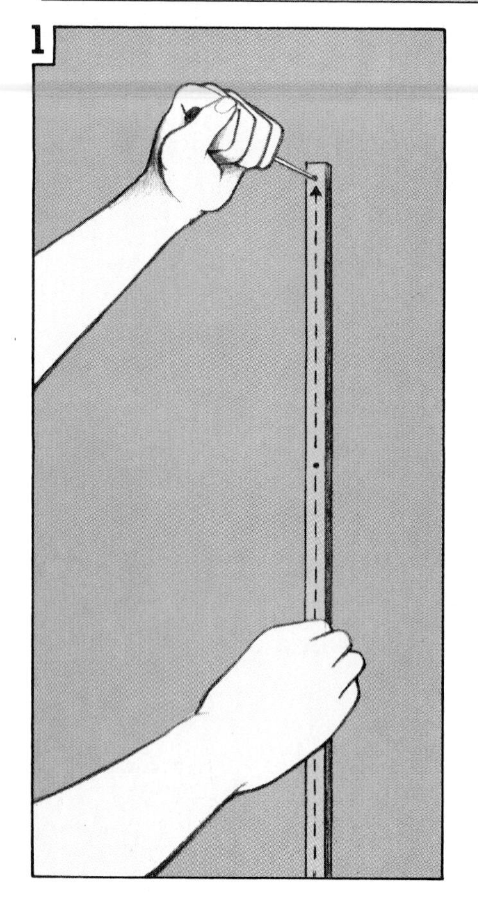

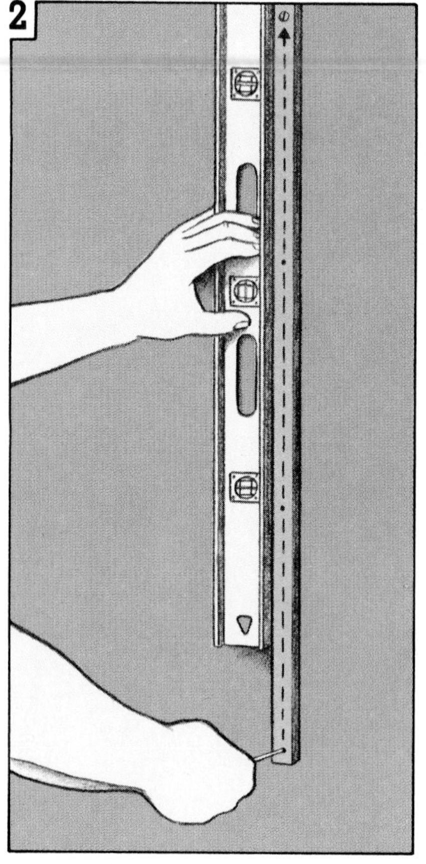

1 When installing shelf standards with horizontal holes, look along their lengths for a manufacturer's label, numbers, or other marks that indicate the standards' tops and bottoms. With this type it doesn't matter if you install the standards right side up or upside down, as long as you place *all of them the same way.* Standards that have vertical holes, however, have no tops and bottoms. If you need to shorten the standards, place them in a vise (the clamped part protected by cloth) and cut with a hacksaw.

Begin the actual installation by holding one of the end strips against the wall at the desired location. Now mark the location of the top hole with an awl as shown and drill a pilot hole.

2 Insert your fastener and partially tighten it. Using a carpenter's level, plumb the dangling strip, then mark and drill for the lower holes. Insert the fasteners for those, then secure the strip firmly against the wall.

3 Next, run a 2×4 straightedge from the top of the strip to the approximate location of the last strip. Strike a light, erasable line along the straightedge as shown in the illustration.

Now step back to see if the line "looks" level. If it does not, redraw the line so that it is parallel to the floor or ceiling, whichever is closer.

Now, measuring out from the first strip, mark on the line where you want the last strip, and install it according to steps 1 and 2.

4 If your shelves need intermediate support strips (check the span table for help determining how many you need), here's an easy way to figure proper spacing. Measure the distance between the centers of the end supports and divide by the number of spaces between the supports (the total number of supports minus 1).

For example, if the overall distance measures 120 inches, and you have a total of four support strips, the equation becomes 120 inches ÷ 3 = 40 inches. In this example, you'd position the intermediate strips at 40 and 80 inches.

5 Though a variety of ways exist for mounting *shelving units* on walls, the method shown here lets you hide a cleat support system while providing plenty of holding power.

First make a bevel cut along the length of a 1×4 (see pages 36-37 for information on beveling. Level and secure one of the member's (beveled edge up and facing toward the wall) to the wall at the desired height and location.

Now secure the other member between the sides and against the top and back of the unit. (The unit's back must be inset ¾ inch.) Lift the unit onto the wall ledger.

Another reliable method—ideal for shelving units whose backs are not recessed—involves attaching 1× cleats horizontally to the top and bottom inside surface of the back, then securing the unit to the wall through the cleats.

3

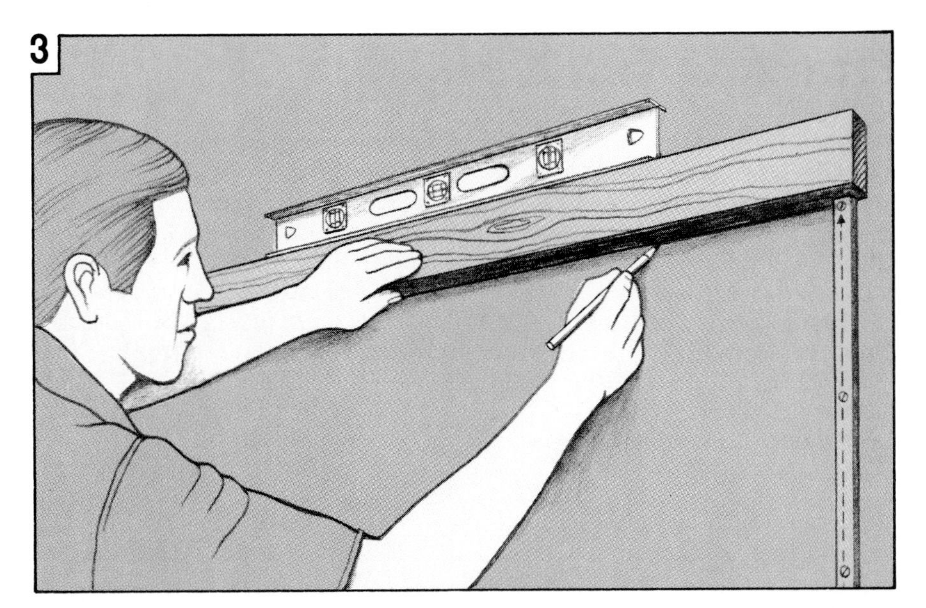

4

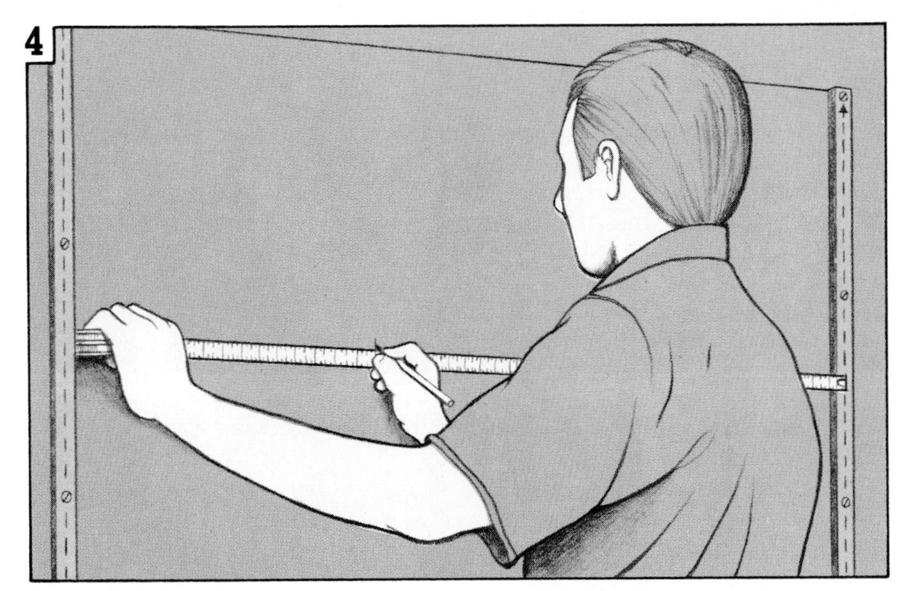

5

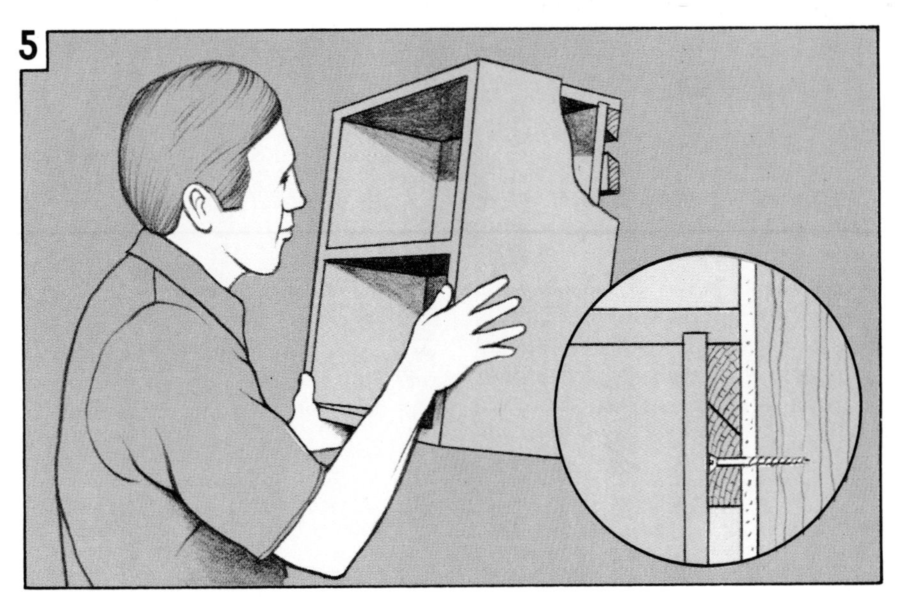

Framing Walls

If you could look at the 2×4 framework behind most finished residential walls, you would discover a rather simple construction. Upright members (called studs) butt against horizontal members (plates) at top and bottom. But building such a wall, whether to finish off a basement or some other room, doesn't just happen. It requires some thoughtful planning and know-how.

You must examine the situation at hand before determining the wall assembly method that's right for you. If the floor and ceiling are nearly level, *preassembly* makes good sense. If, on the other hand, the wall and ceiling are uneven, opt for *building a wall in place* (see page 77). With the first method you build the wall on the floor, then raise it; with the latter, you custom-cut each stud to fit and nail it to top and bottom plates already in place.

Whichever way you go, try to position the wall perpendicular to or directly beneath the ceiling joists so you have a sturdy surface to attach the wall to.

Preassembling on the Floor

1 Begin by deciding on the wall's location. Then, lay a framing square against an existing wall as shown, and have a helper hold one end of a chalk line against the outside corner of the square while you extend the line's other end along the square's blade to the desired wall length. Pull the line taut, reach in toward the middle of the line, and snap it.

2 Now you're ready to secure the nailer plate, which, though not ab-

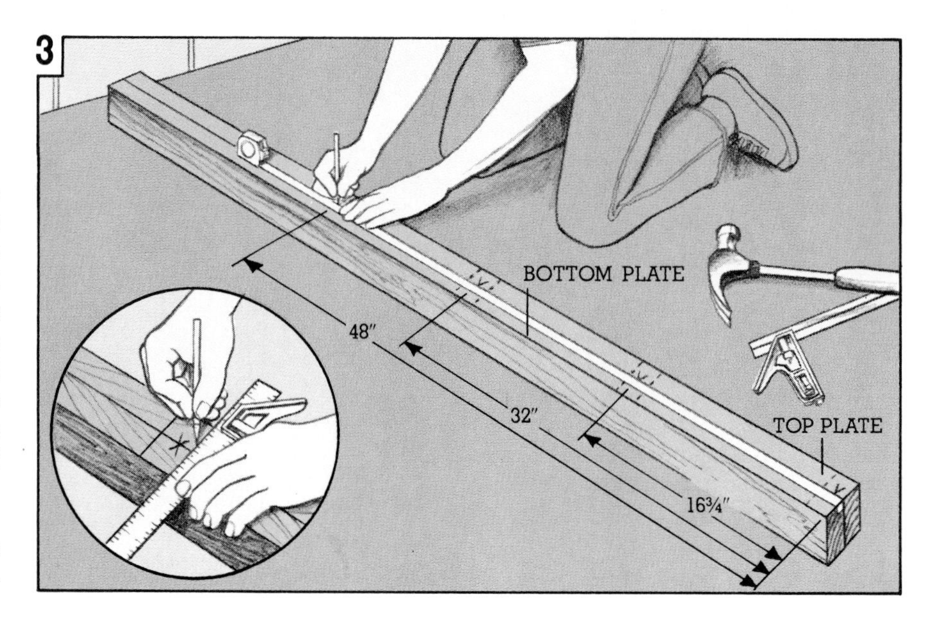

solutely necessary, provides the firmest possible hold on a concrete floor for preassembled walls. Cut a 2×4 to size, align it with the chalk line, and drive concrete nails through the plate and into the floor. (For long walls, cut additional 2×4s to the needed length and butt them end to end before securing them.) To make nailing easier, drill pilot holes through the 2×4s and on into the concrete using a carbide-tipped bit.

3 Cut the bottom and top plates, and mark stud locations on them as shown here. Place the plates on edge alongside of one another and mark the center of the first stud location ¾ inch in from the end. Now hook your tape on the end of the plates and extend the blade down along the members. Place stud center marks at 16¾ inches, 32 inches, 48 inches, and every 16 inches thereafter. Then, using your center marks as a reference, and a framing square, mark the studs' outlines.

4 To obtain stud lengths for the wall, measure the distance from the bottom of the joists to the top of your nailer plate at several points along the proposed wall. Subtract 3 inches from the shortest distance (to account for the top and bottom plates), and cut all studs to that dimension. Since the wall may only touch the ceiling joists at one location, you can expect to do some shimming.

5 If you weren't able to position the wall below or perpendicular to the floor joists, glue and nail blocking between the joists as shown here. Ideally, you should use the same size material your joists are made of, usually 2×8s. Space the blocking 16 to 20 inches apart to provide plenty of support.

6 Working on a flat surface, lay the studs on edge between the top and bottom plates. It helps to have something solid, such as a block

(continued)

Preassembling *(continued)*

wall or the nailer plate, to nail against while assembling the wall.

For speed, nail one plate at a time to the studs. Drive two 16-penny common nails into each end of each stud. Since hammer blows tend to knock studs out of alignment, continually double-check your work while nailing. Keep the studs' edges flush with the plates' edges, and if any of your 2×s are twisted or bowed, replace them. If you don't, these deformities will detract from the looks of the finished product.

7 Now comes the fun part—raising the wall! Try to have a helper on hand, especially for lengthy walls. The framework you've assembled is both heavy and bulky. Begin by resting the bottom plate against the nailer plate, then tip the wall into position. If the wall won't quite clear the joists, tap the top and bottom plates alternately with your hammer until the edges of the bottom plate flush up with those of the bottom nailer plate, and the top plate appears roughly straight up and down.

8 If, on the other hand, the wall is a bit short, drive shims between the two bottom plates as shown. Have your helper steady the framework while you drive the shims. (Remember to shim both sides to balance out the load of the structure and to prevent it from tilting.)

9 Once the wall is snug, nail the bottom plate to the nailer plate with 10-penny common nails, angling them for the best possible hold. Check the framework for plumb, in two directions, with a level. If adjustments are needed, take the level away and tap the framework into position. Check for plumb again, then go ahead and nail the top plate to the joists.

Building a Wall in Place

By building a wall in place, you not only eliminate one of two bottom plates found in preassembling, you also sidestep having to shim.

1 Start by cutting the top and bottom plates the desired length, then marking stud locations (see page 74). Though spacing is the same used in preassembly, place marks here on the plates' faces for effective alignment while toenailing. Now nail the top plate to the joists as shown here. You'll need a helper to hold the plate against the joists while you nail. And don't forget to wear protective goggles while hammering.

2 Next, have your helper dangle a plumb bob from the end of the top plate and at two or three points along one edge of it. Or if you're working alone, dangle the bob from a nail driven into the edge of the top plate. Mark the exact location of the plumb bob on the floor. Connect the marks by penciling a line along a straightedge or by snapping a chalkline. Secure the bottom plate to the floor as explained on page 75.

3 With both top and bottom plates installed, measure each stud length. Add 1/16 inch for a snug fit, then cut the studs. Tap each stud into place.

4 To secure the studs, drive nails at a 45-degree angle through the end of a stud and into the plate as in the sketch on page 78. Don't worry if the first nail you drive moves the stud. The second nail, *(continued)*

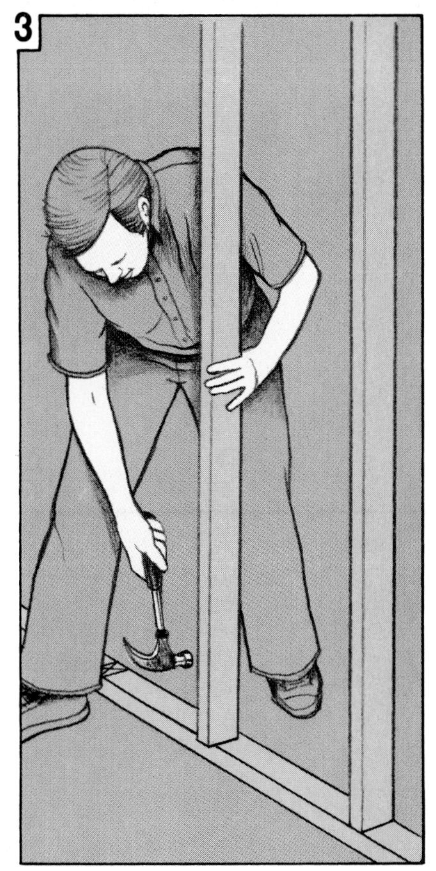

Building a Wall in Place *(continued)*

driven in from the other side, should drive it back. For extra strength, drive in a third nail down through the edge.

If the stud won't line up, try knocking it with a baby sledge. Or cut a spacer the appropriate length and place it between the last secured stud and the one you're about to install. Still another way would be to drill holes at an angle through the stud first. This eliminates the shifting.

Framing Corners and Intersecting Walls

If your project calls for building two or more walls that connect to each other, you need to know how to join the wall sections together properly. This involves making sure the walls are square with each other and providing a nailing surface for the wall material that will later dress the framework.

Here, we show three ways to get the job done. In situation 1, the extra stud offers a nailing surface for finish materials and buttresses the corner as well. To tie two walls together using this framing arrangement, drive 16-penny common nails through the extra stud and into the end stud of the adjoining wall, clinching the nails for a firm grip (see page 47.)

In situation 2, several foot-long 2×4 scraps (use three in a standard 8-foot high wall) serve as spacers between two full-length studs placed at the end of one wall. Tie the wall sections together with 16-penny common nails.

Situation 3 shows what the framing should look like when two walls intersect. Nail three studs together and to the plates at the point of intersection. Then center the intersecting wall between the two outside faces of the three studs, square the corners, and nail the wall into place.

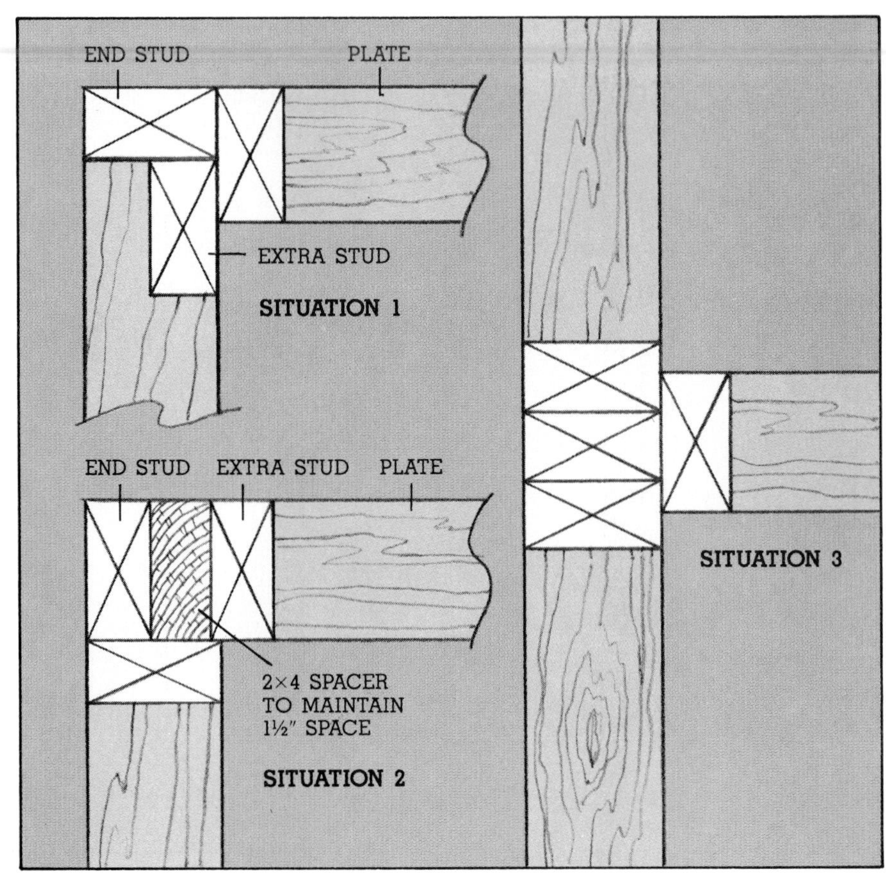

Roughing-In an Opening

If your wall-building plans include a door opening, but you're a little unclear about how to do the framing, the information on this page should help you.

To begin with, find out from your supplier the rough opening dimensions for the size and type of door you want. Normally, hinge-type doors come in 18-, 20-, 24-, 26-, 28-, 30-, 32-, 34-, and 36-inch widths; door heights remain standard at 6 feet 8 inches. Another way to find the needed dimensions is to add 2½ inches to the door width (to allow for the side jambs and shims), and 2 inches to the door height (to allow for the head jamb, shims, and carpeting or other flooring that goes beneath the door).

Once you know the size of your opening, build the wall as described on pages 74-78, only this time include the rough opening members shown below.

Trimmers (vertical 2×4s found at each side of the opening and attached to a stud or another trimmer) provide solid, unbending support for the door that will hang from them.

The *header* (two 2×6s with a ½-inch plywood spacer sandwiched in between) spans the top of the opening, providing a rigid defense against bowing that results from a door's weight or any overhead load. (For openings less than 3 feet wide, replace the 2×6s with 2×4s placed on edge.)

Cripples (short lengths of 2×4s between the header and the top plate) help maintain a 16-inch stud spacing for nailing on sheet goods, and distribute the weight equally from above.

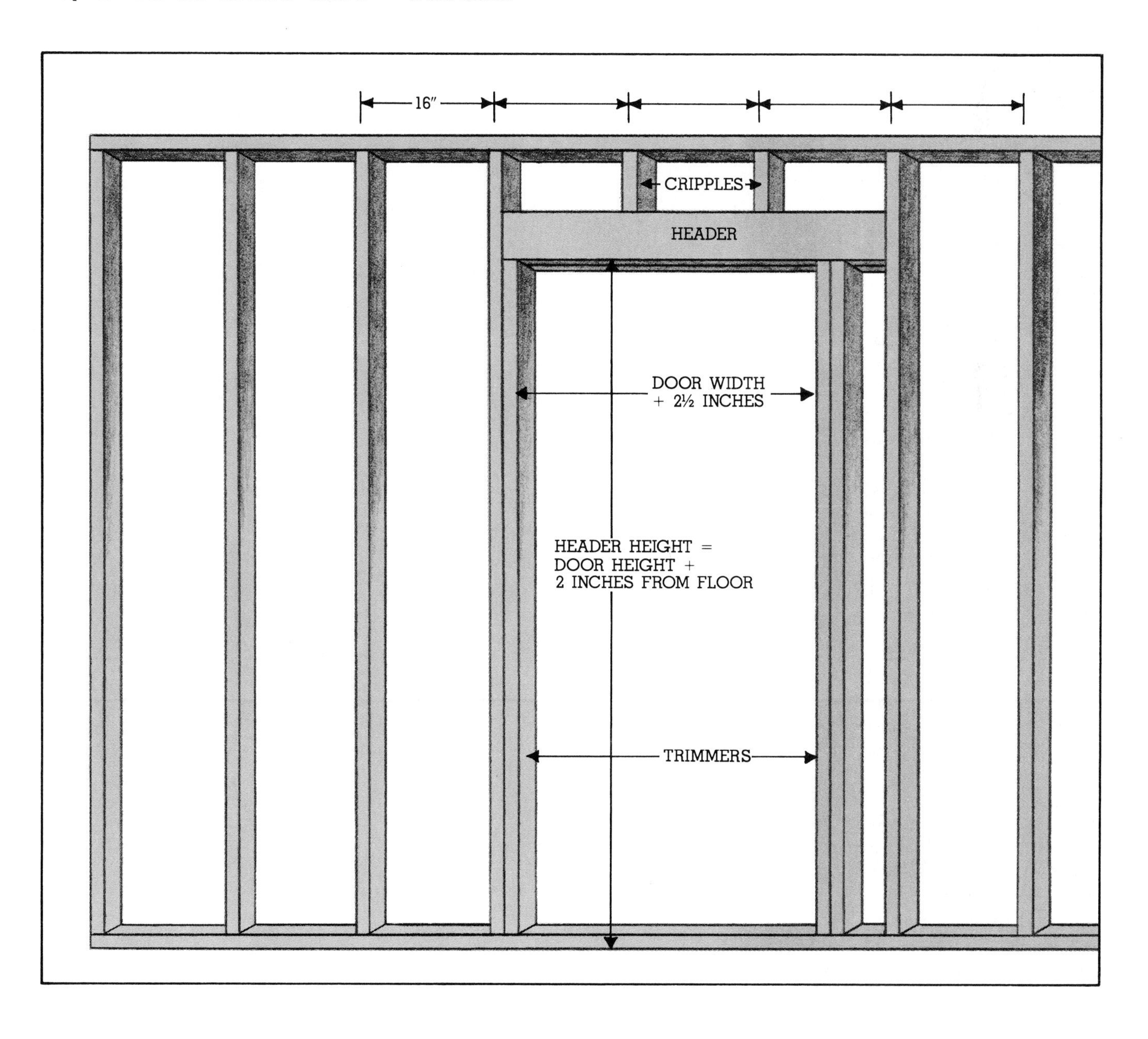

Other Framing Situations

Though there's no substitute for experience in carpentry when confronted with a new framing situation, using ready-made solutions such as those shown here can save you lots of time and trouble.

1 For framing-in a soil stack, rely on the chase construction below. Measure, cut, and preassemble three wall sections that consist of short top and bottom plates and two studs. Placing 2×4 cross members at 2-foot intervals between the studs further solidifies the chase. Now raise and secure each section to the floor, ceiling, and adjoining framing.

2 To frame-in a steel I-beam or a triple 2×12 beam in a basement, follow this procedure. Hang horizontal 2×4 rails along the length of the beam from the ceiling joists via notched 2×4 vertical supports. Tie the sides together and ensure consistent spacing with 2×2s of equal length. If you are fortunate enough to have a helper, build the framework on the floor, then lift and nail it in place.

3 Another familiar framing situation, the closet may look complicated, but it's really not. It's just a short wall with an opening (see page 79) joined at a typically framed corner (see page 78) to a shorter wall with no opening. Note that the header is simply a pair of 2×4s turned on edge and sandwiching a length of ½-inch plywood. You can use 2×4s no matter how wide the closet opening.

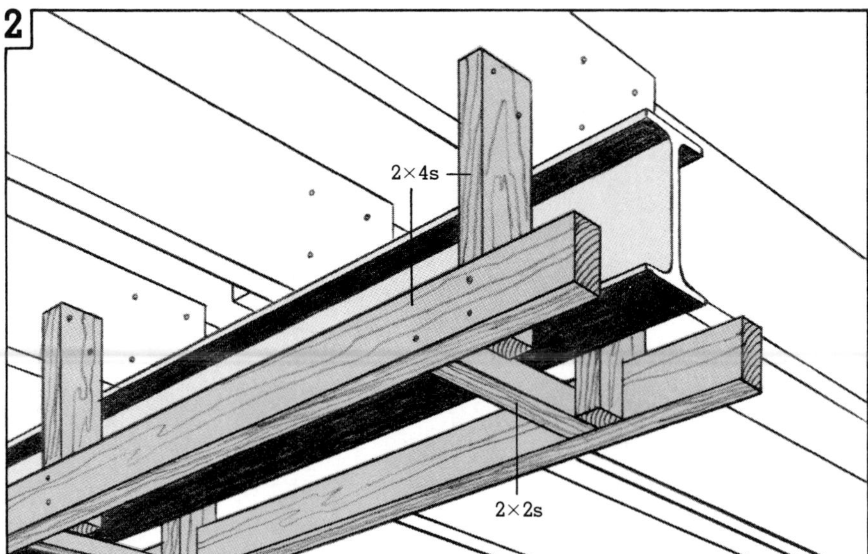

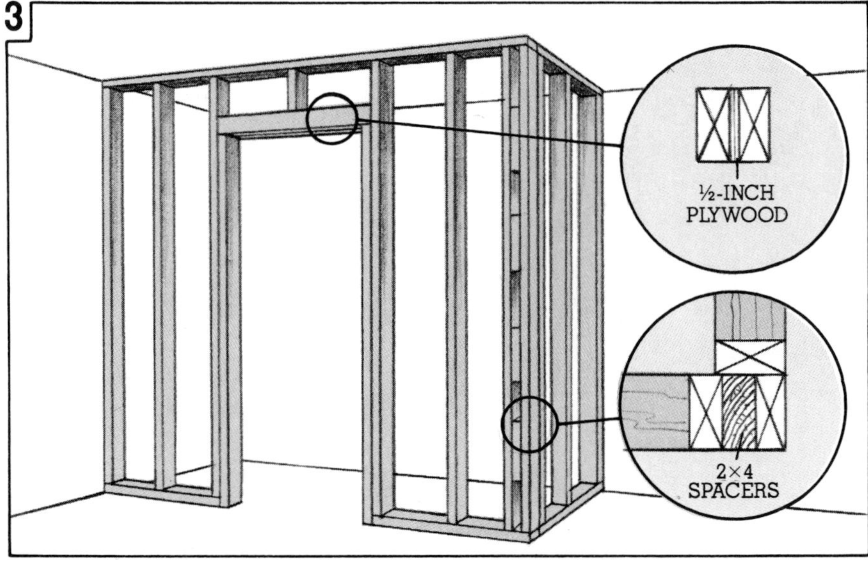

Furring Out Walls

On pages 74-80 we take a thorough look at how to build stud walls. Here we discuss an alternative. Furring out a wall (see the typical layout below) is especially appropriate in those situations in which you can't afford to sacrifice the room stud walls require. But 1×2 or 1×3 furring strips have more going for them than being able to save you valuable space. They're also easier to work with and less expensive than studs.

To find out what's involved in readying a wall for paneling, drywall, or other material requiring a solid, plumb backing, turn the page.

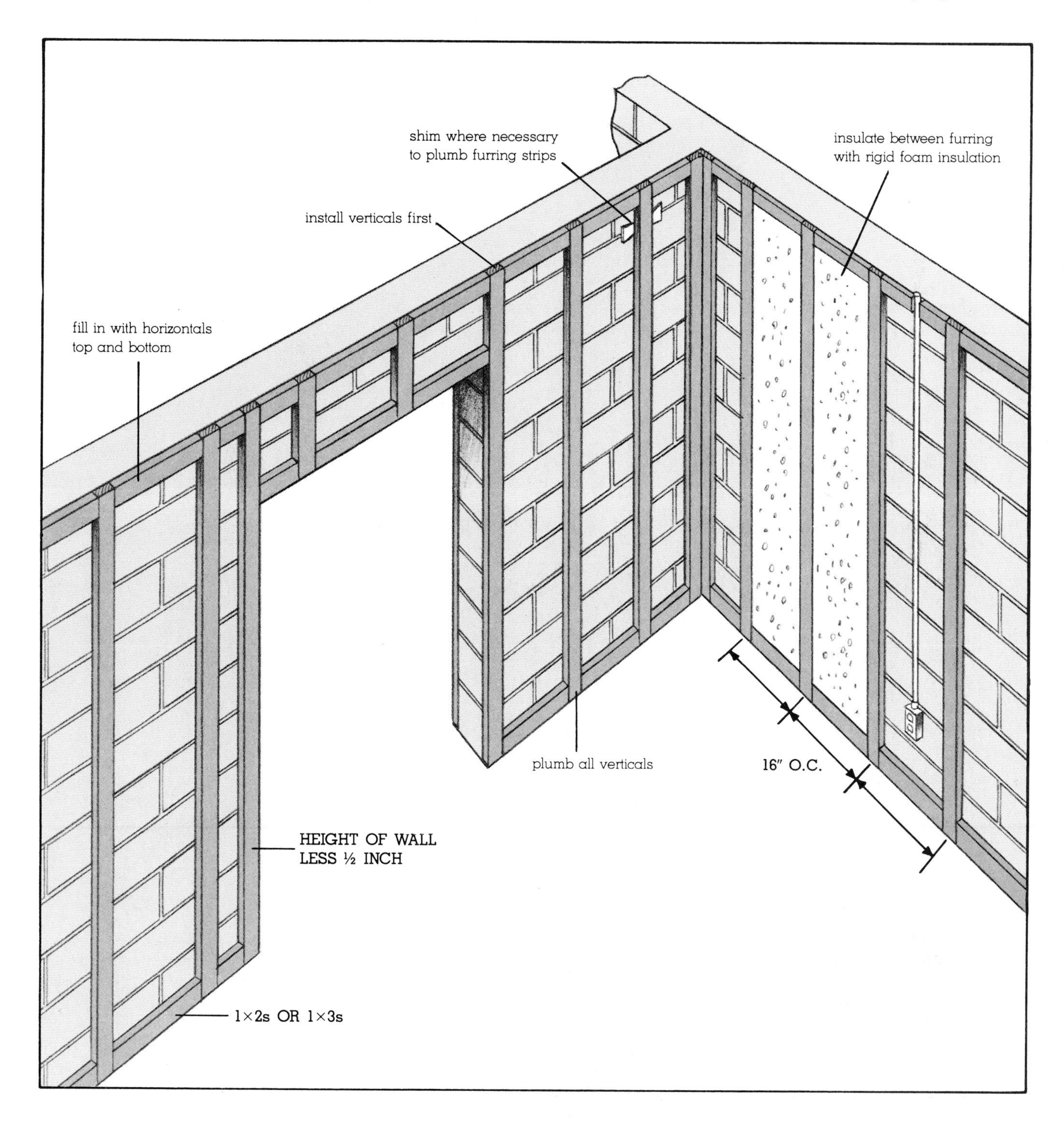

Furring Out Walls *(continued)*

1 Start by marking the vertical furring strip locations. One easy way to do this is to position a sheet of the desired wall material in the corner of the room, plumb it, and strike a line down along its outside edge. Using the line as a guide, and 16 inches as your center-to-center measurement, mark the location of the other vertical strips along that wall.

Now, measure and cut each strip (one at a time) to fit between the floor and ceiling. (If yours is a basement construction, subtract ¼ inch from each measurement—the distance you will want your wall to be above the floor as a safeguard against flooding and settling.)

Next, apply a wavy ¼-inch bead of panel adhesive down one side of the furring strip as shown in this sketch. Now raise and center the strip on the line, pressing it firmly into place to help spread the adhesive.

2 Pull the strip back off the wall and prop it against the back of a chair or the wall. Doing this helps the adhesive dry more quickly. After the amount of time that is specified on the label of the product you're using), again press the strip into place.

3 Using your carpenter's level, plumb the furring strip as shown in the sketch. Double-check the 16-inch spacing, too. (If any adhesive makes contact with and sticks to your level, be sure to wipe it off immediately with a rag dampened in mineral spirits.)

4,5 Now check the strip for plumb as shown. If all is well, secure the vertical strips to the wall with common nails (for wood frame walls) or specially hardened masonry nails (for concrete or masonry walls). If, however, some glaring irregularities exist, which you can discover by holding a straight 2×4 perpendicular to the wall after installing the furring, drive pairs of shims (wherever necessary) as shown behind the strips.

6 Drive the fastener through the strip and shims, and into the existing wall (into studs or mortar joints wherever possible). Do the same to all remaining strips. If going into a masonry wall, you'll be best off driving the masonry nails with a baby sledge.

7 After installing all the vertical strips, begin work on the horizontals. For these, measure between the vertical strips at the top and bottom. Apply adhesive, shim, if necessary, and install as already described.

3
4
5
6
7

Installing Drywall

Drywall is a do-it-yourself dream material. It is easy to handle, cut, and hang; doesn't cost much; and covers lots of territory fast. Below and on the following pages we tell and show you how to hang and finish drywall like the pros do it.

Before actually hanging the drywall, however, study the sketch below. It gives the drywalling rules of thumb that will save you time and trouble. Then, make sure you have a sturdy backing of stud walls or furring strips on which to hang the drywall. For information on both of these, see pages 74-83. And if you plan to insulate, or do any plumbing or electrical work, do it now. Be sure to use a vapor barrier to protect against moisture.

And lastly, determine your drywall and other material needs. To do this, first multiply the *length* by the *width* of each wall or ceiling to find the number of *square feet* to be covered. From this number subtract the square footage of all door and window openings, then add 10 percent to the remainder for waste allowance. Now divide this figure by the number of square feet in each sheet.

You'll also need one 5-gallon container of premixed joint compound, a 250-foot roll of joint tape, and 5 pounds of 1½-inch ring-shank nails for each 500 square feet of surface. For outside corners, order *corner bead.*

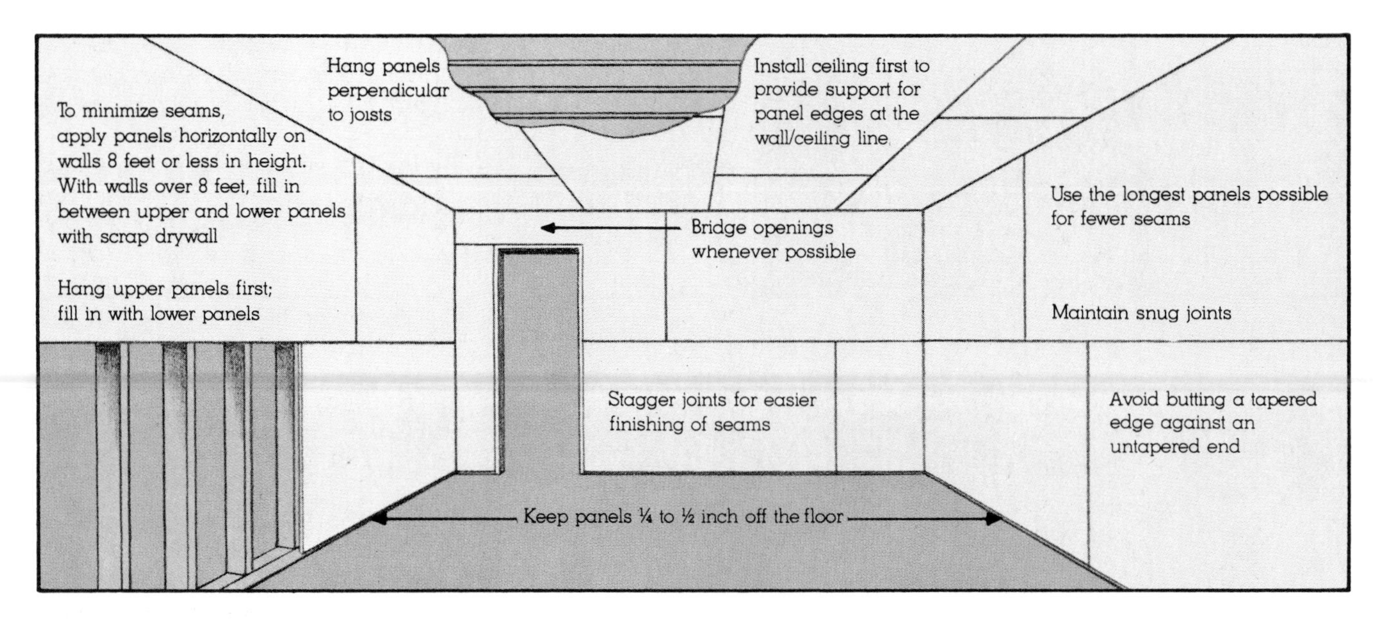

Measuring and Cutting

When cutting drywall to fit, the pieces should rest snugly against their neighbors, with measuring and cutting errors not exceeding ¼ inch. This is especially true for corners that are out of square, electrical outlets, and pipes, all of which require custom-made pieces.

1 Most generally, it's best to cut drywall on the floor. After making sure there aren't any stray nails, drywall scraps, or other objects

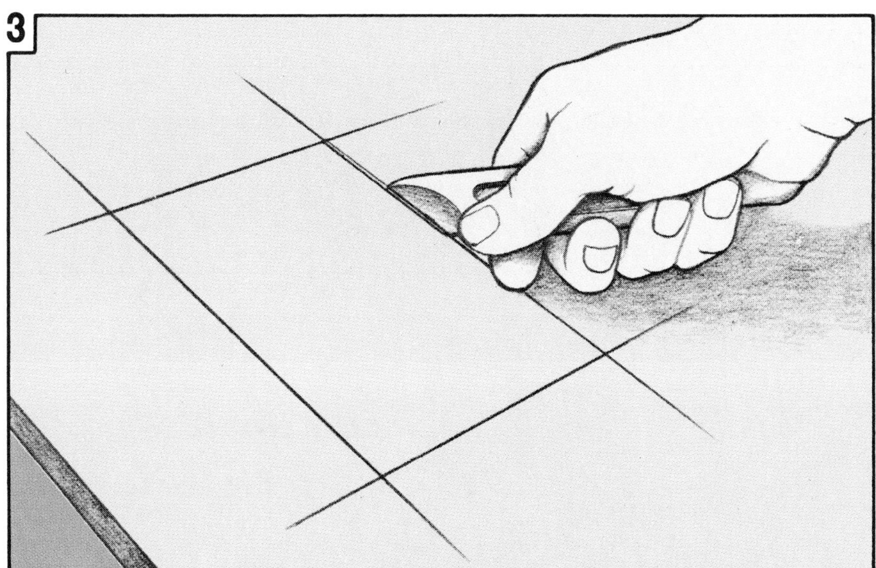

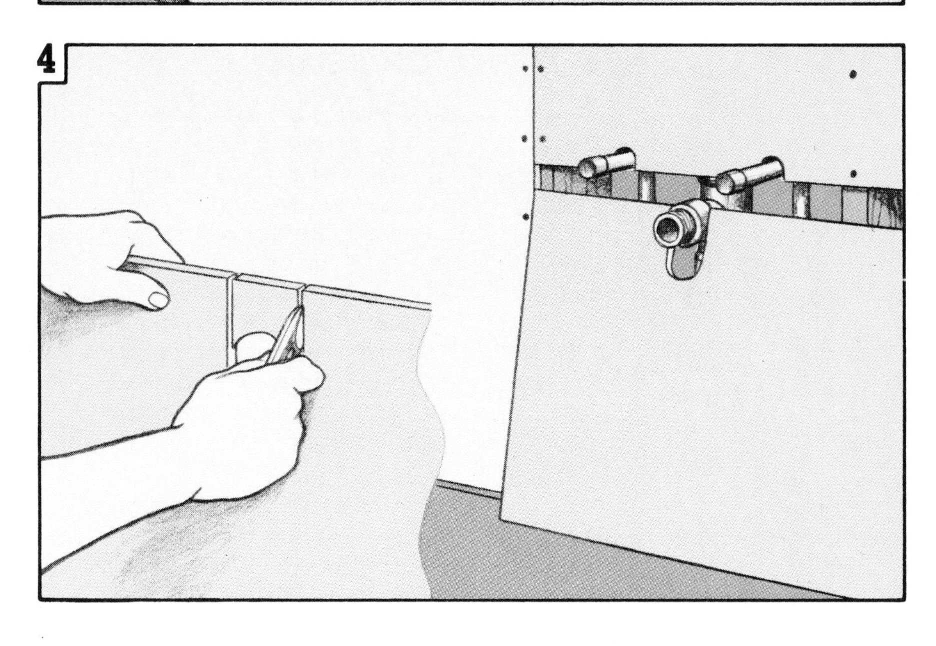

beneath the panel, press a straightedge firmly against the cutoff marks. Now, score along the cut line with a utility knife, severing the top layer of paper and slicing into the gypsum core. Next, lift the panel onto its edge, snap the cut segment back and cut the backing still connecting the two pieces.

2 If you need only to determine the cutoff height of a panel, measure the distance from the floor to the ceiling at the panel's left and right edges, and subtract ¼ inch from the shorter distance.

To determine the correct cutoff width of a corner panel, measure the distance from the last panel at the top and bottom to the corner. Transfer these markings to the panel, then make your cut.

3 To cut around an outlet box, measure the distance from the box's edges to the edge of the last panel. Then measure the distance from the box's top and bottom to the floor minus ¼ inch. Transfer these onto the face of your panel, extend the lines from the marks until you've outlined the box, then make the cut, using a utility knife or drywall saw.

4 A different kind of surgery awaits you when cutting around plumbing pipes. First, mark and cut out the needed holes in a piece of drywall, using a hole saw. Then, cut the panel into two sections so you can slip them around the pipes for a snug fit.

Hanging the Panels

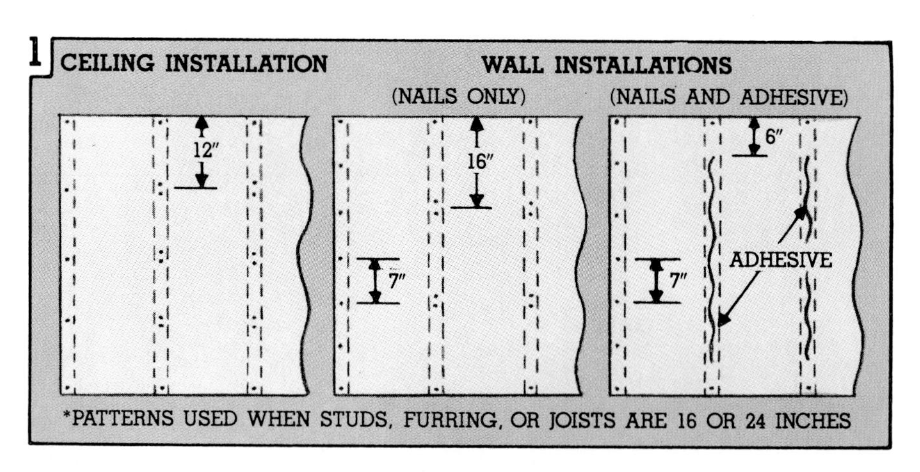

1 Here are three *effective* drywall fastening patterns. Note that with ceiling panels, you should drive nails at 12-inch intervals around the perimeter, and double-nail every 12 inches along each joist. Why double-nail? Even if one of each pair pops, the other will keep the panel from sagging. (Always use ring-shank nails to secure the panels; 1½ inchers with ½-inch drywall, 1⅝-inch ones with ⅝-inch material.)

When hanging wall panels, you have two ways to go. If you use only nails, drive one every 7 inches around the border, and double-nail every 16 inches along each stud. If, however, you use adhesive and nails, run a bead of adhesive along each stud the panel will cover, except the two at its borders, making sure to keep it at least 6 inches from the panel's edges. Then position the panel and drive nails at 7-inch intervals along the border. If the panel bows, drive one nail into each intermediate stud or furring strip midway between the panel's ends or edges.

(Note: Drive perimeter nails ⅜ inch in from the panel's edges.)

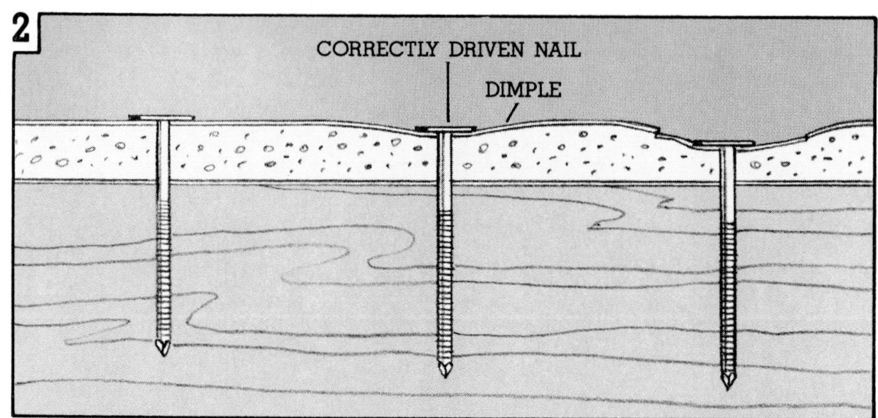

2 As you can see here, there are several ways to drive a drywall nail—but only one correct one. Driving the nail flush will cause problems when you try to conceal it with compound and texture. If you drive the nail too deep and tear the cardboard facing, the nail won't hold the panel to its backing.

But if you manage to just barely dimple the drywall with the last hammer blow, neither of these problems will occur.

3 To install ceiling panels, start in a corner and against one side of the room and work out from there, keeping panels perpendicular to the joists. When placing a panel, try supporting one end of it with

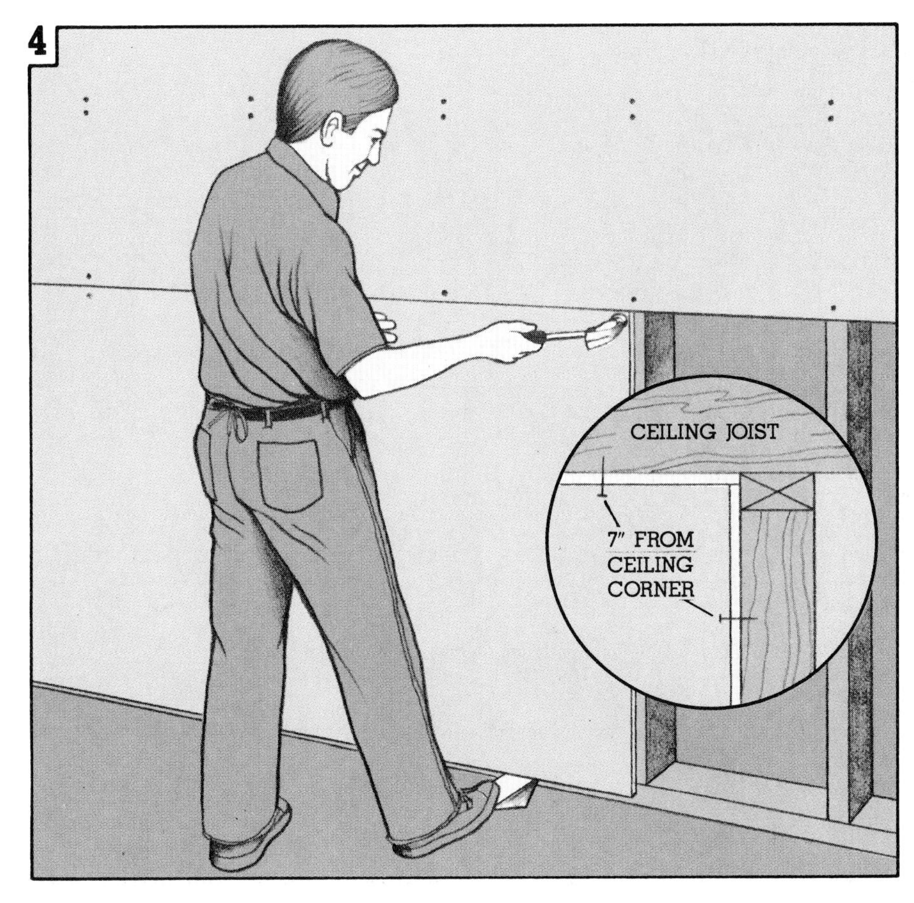

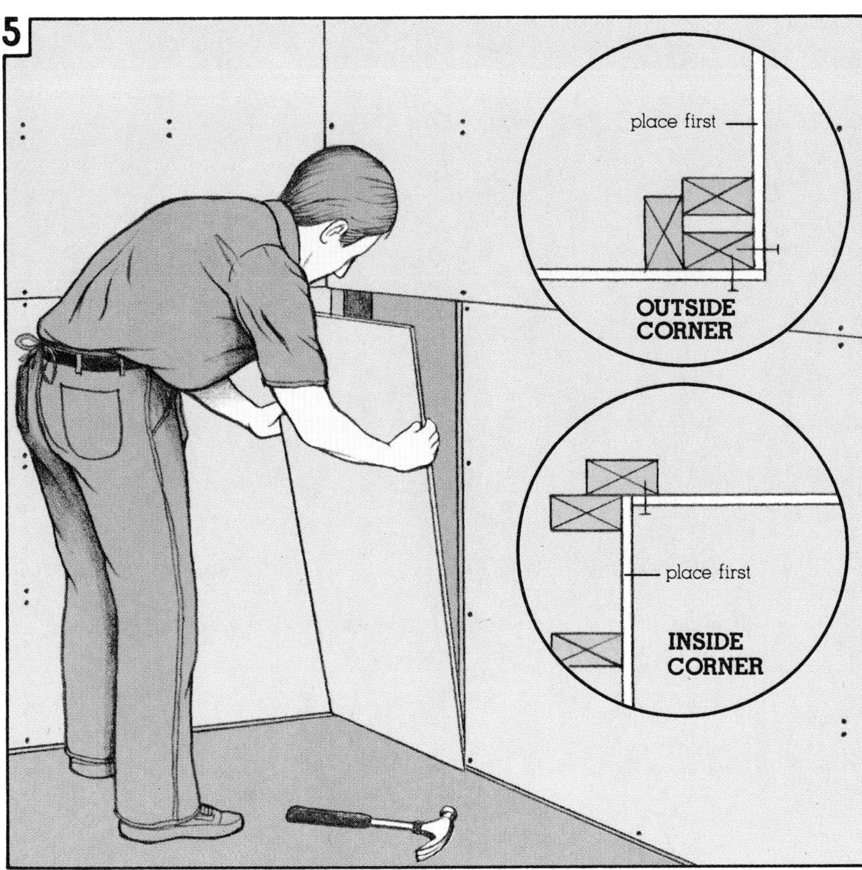

your head, leaving your hands free to hold and drive nails. Have a helper support the other end. When nailing, drive nails around the panel's perimeter first, then fill in between as shown in sketch 1. If you're working alone, consider renting a special drywall lifter like the one in the detail.

4 Once the ceiling panels are in place, begin work on the walls, installing the upper panels first. (It's a good idea to start a few nails into the drywall before lifting the panel into place.) Make sure those panels butt firmly against the ceiling before nailing. As shown in the detail, keep nails that are driven into the upper panels at least 7 inches from the ceiling corner. If the end of a wall panel fails to fall midway across a nailing member, trim it with a straightedge and utility knife so that it does.

When installing the lower wall panels, fit them firmly against the upper panels—tapered edge to tapered edge—raising them with a wooden wedge as shown. Remember, too, to maintain a ¼- to ½-inch spacing along the floor.

5 You'll probably have to cut filler pieces to finish any drywall job. To do this, simply measure and trim each piece to size, making sure that it has at least two nailing members to support it. Now insert the pieces—cut edges against the corner.

Taping and Texturing

There are no two ways about it: Taping drywall seams and corners is an art. Even some professionals can't do an adequate job of taping every time. But given the guidelines described here, and with some care and practice, you should end up with a wall that looks professionally done.

Naturally, the more you practice, the better you'll get. So if you're just beginning, work on some scrap drywall first. And when you do begin taping the real thing, start in an inconspicuous area—a closet would be great.

In most cases, three coats of compound will give satisfactory results. Let each coat fully dry for 24 hours before spreading another. (Humid weather, however, can extend the normal drying time.) Note: You can buy a 90-minute quick-set joint compound from some drywall suppliers, but unless you are exceptionally speedy you needn't bother with it.

Once taping is complete, texture the ceiling and walls, in that order, to disguise any imperfections that remain. You can blow-on texture materials of various kinds with a rented texture gun and hopper, or roll them on using a paint roller.

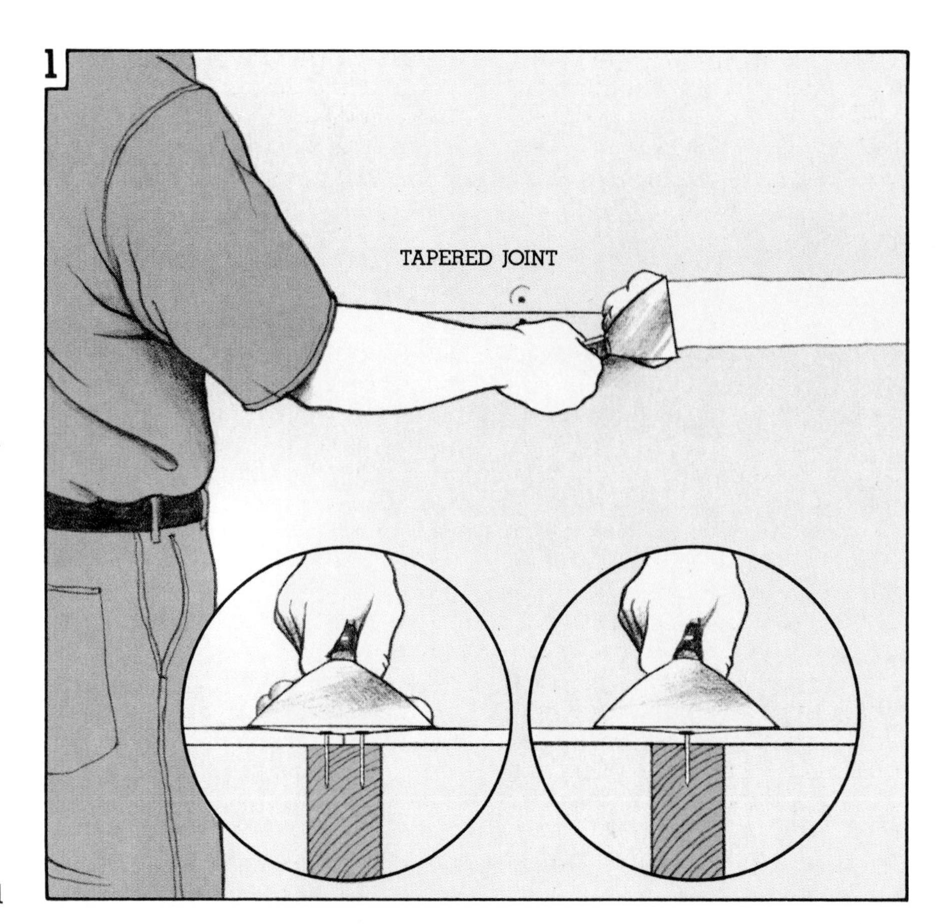

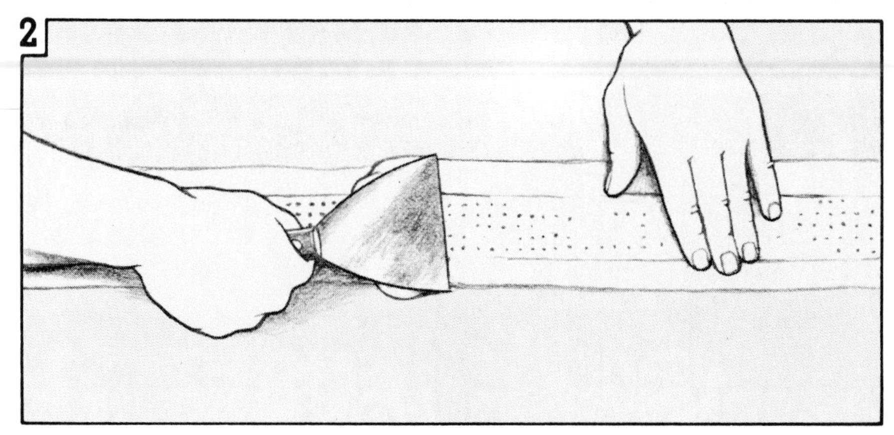

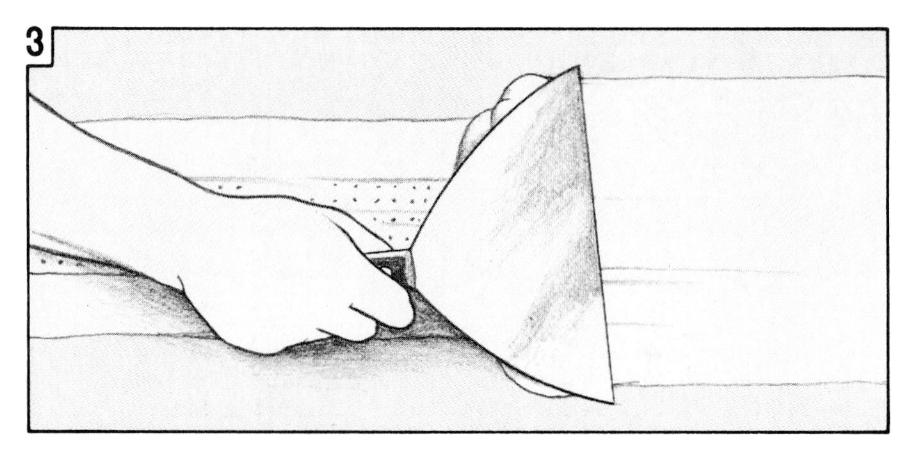

1 Before applying compound to tapered wall or ceiling joints, make sure adjoining panels are flush with one another. Then, using a 4- to 6-inch-wide drywall knife held at a 45-degree angle to the wall surface, spread an even *bed coat* of compound over the joint depression as shown.

If you're working with an end joint, spread a thin layer over it, being careful not to build up a noticeable mound that later would require extensive feathering and sanding.

To conceal nailheads, pass over them several times—until the depression fills completely with drywall compound.

2 Immediately after applying the bed coat to a joint, center a length of drywall tape over the joint and press the tape firmly against the filled joint with your hands. Then, using one hand to hold the tape, further embed the tape by pressing it (but not too hard) with a taping blade. If bubbles form beneath the tape, peel it off and apply more compound; then press the tape back against the joint. Do the same, too, if wrinkles occur.

3 After the bed coat dries, load a 10-inch-wide drywall knife with compound and apply a topping coat over the joints. Feather out the material to about 6 or 7 inches on each side of the joints in order to blend the compound in with the surrounding surface.

For end joints, feather out 7 to 9 inches on each side; for nail dimples, apply another coat to fill the shrinkage cracks of the first coat.

In all cases, apply additional skim coats over the second coat if needed. Then, once the final coat dries, smooth the surface using 80- or 100-grit open coat sandpaper or a dampened sponge.

4 With outside corners, you have a slightly different operation. To protect and conceal the raw drywall edges that meet at an angle, cut a strip of corner bead the appropriate length, using tin snips. Fit the strip over the corner, then fasten it to the wall one side at a time. Drive nails at 10-inch intervals through the nail holes on each side of the strip and across from one another.

Next, apply a bed coat of compound over the corner bead with a 4-inch knife angled away from the corner. Allow one side of the blade to ride on the bead; the other side on the wall. Spread the compound 3 to 4 inches on each side of the bead's nose until the depressions underneath are filled and smooth. Wipe up any excess and let dry.

Later apply a second coat, feathering the edges 7 to 9 inches from the corner with a 10-inch knife. Add another coat(s) if needed, let dry, and sand or sponge.

5 For inside corners, begin by applying a bed coat, first down one side, then the other, with a taping knife. Cut a piece of joint tape to the correct length, fold it, then insert it by hand as shown here. With your knife angled away from the corner, wipe down each side, applying enough pressure to embed the tape. Let the compound dry. Apply another coat (and a third one if necessary) to further feather the compound. After the final coat dries, sand or sponge the surface smooth.

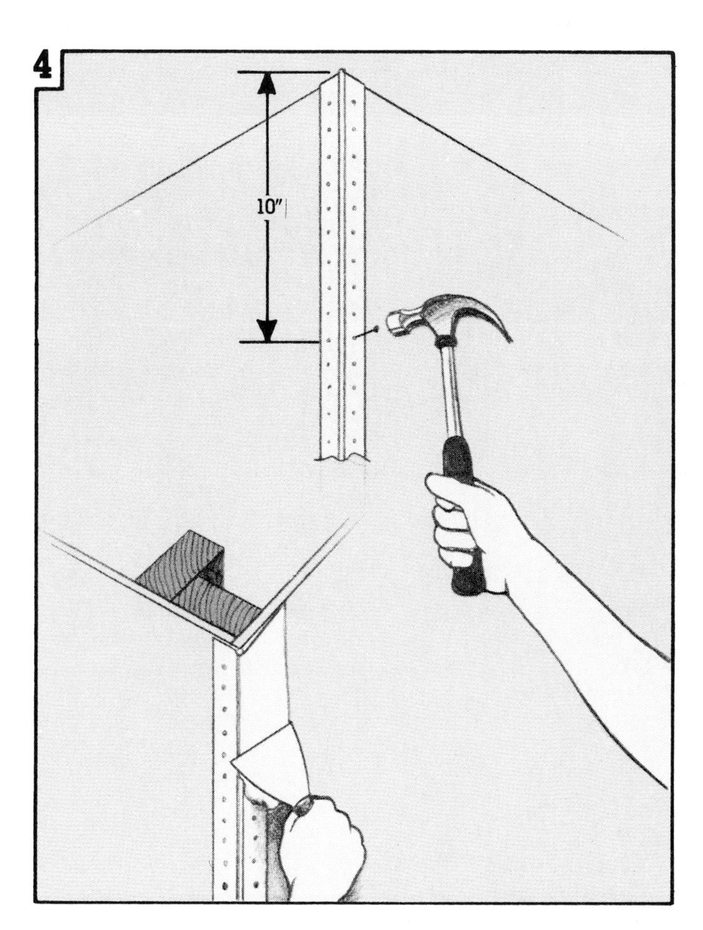

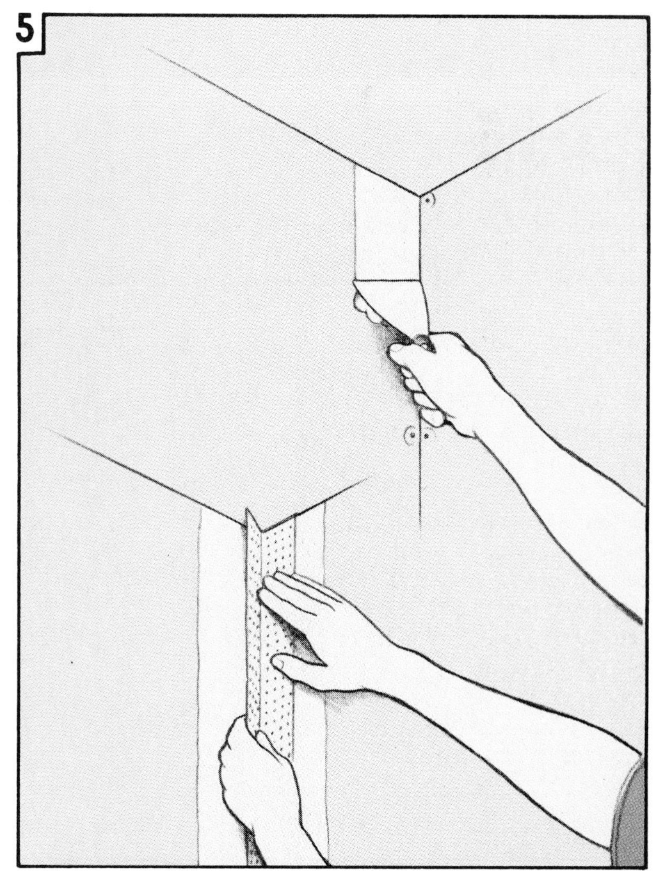

Installing Paneling

Since drywall covers most interior walls, it tends to take on a drab, everyday look. Easy-to-install paneling, however, offers an exciting contrast and beautifies walls with the rich look of solid wood.

As with drywall, which we discuss on pages 84-89, paneling needs a solid, plumb backing to hang properly. Typically, this is a stud wall or furring strips fronted by ½-inch drywall. On outside walls you should sandwich insulation and a 2-mil polyethylene vapor barrier between the studs and drywall to protect the panels from moisture damage. In a below-grade installation where a chronic dampness problem exists, you may have to keep a dehumidifier going as well.

Estimate paneling needs as you would for drywall (see page 84). And after you bring the panels home, stand them up for 48 hours in the room in which they'll hang to condition them to their new surroundings.

1 When installing paneling over drywall, first locate the studs where the panels' edges will meet. Using a color paint that matches the panels' seams, paint the drywall directly over these studs. This will hide the 1/32-inch spacing you should maintain between panels to allow for expansion.

Then, starting at a corner, measure the wall heights where you expect the edges of the panel to go (minus ¼ inch), transfer the measurements to the panel, then cut the material to fit. Hold the panel up in the corner and plumb it as shown, making sure the outside edge rests in the center of a furring strip or stud. (If the corner is out of square, trim the panel to fit as in sketch 1, page 91.) Strike a line along the panel's outside edge and remove the panel. Apply a ¼- to ½-inch bead of panel adhesive on the panel's backside as shown, then reposition the panel. Use a ¼-inch scrap to elevate the panel the appropriate distance above the floor.

2 After aligning the panel to the plumb line, drive three or more finish or color-matched paneling nails halfway in along the top edge of the panel. With the panel dangling, compress the adhesive behind it by hammering a block of wood wrapped in cloth over the surface.

3 Now pull the bottom of the panel away from the wall, and insert a 4- to 6-inch piece of scrap to keep it there. This helps the adhesive dry and ensures a good bond. After three minutes, remove the scraps, press the panel against the wall, and drive nails every 8 inches along the edge and every 12 inches along the intermediate grooves over framing.

Paneling Around Corners and Obstructions

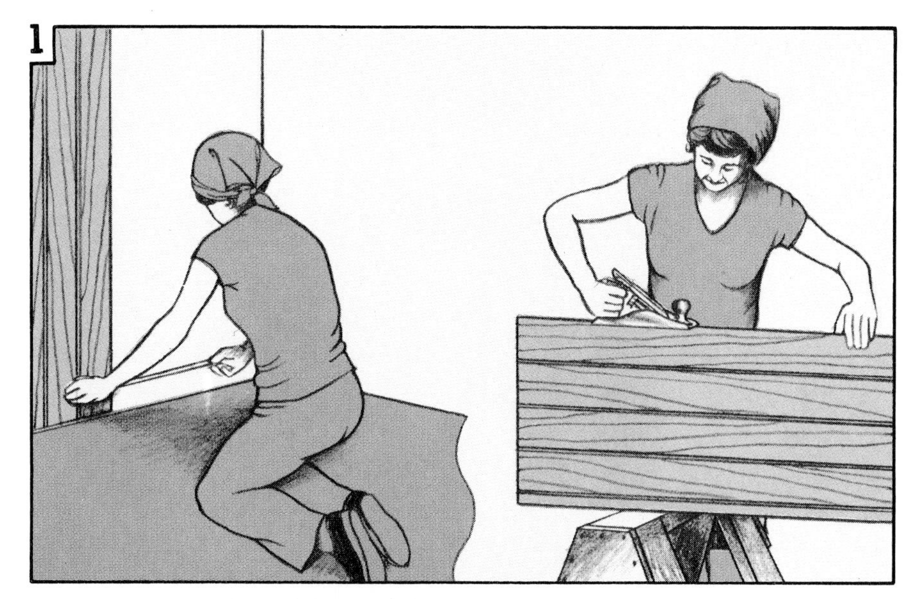

1 Eventually you'll panel your way to an adjacent corner and need a precision-cut panel to finish the wall. To do this, measure the distance from the last panel installed to the corner at several points along the panel's edge. Then, using a straightedge and circular or saber saw with a fine-toothed blade, cut the new panel to fit, keeping the finished side down while sawing to avoid splintering the panel's surface. (With a hand saw, cut with the finished side up.) If irregularities in the corner make it necessary, plane the edge as needed. Then install the panel, maintaining a 1/32-inch space along the seams.

2 To make inside cuts in paneling, as in this case for an electrical outlet, measure the distance between the edge of the last panel installed and the right and left edges of the outlet. Then measure the distance from the floor to the outlet's top and bottom, subtracting ¼ inch.

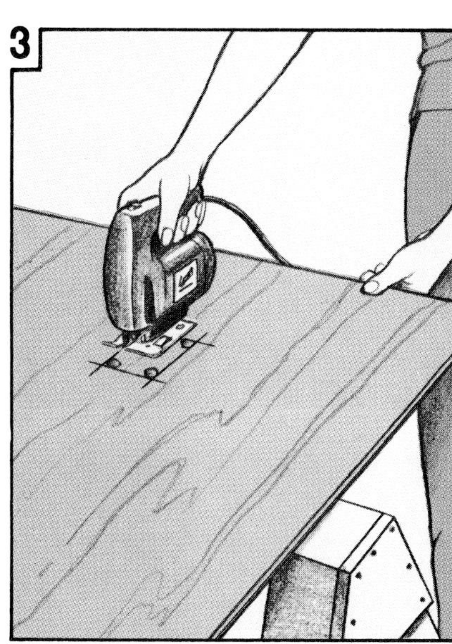

3 Transfer all measurements to the front of the next panel to be installed and outline the outlet. Begin making the needed opening by drilling ¾-inch starter holes in all four corners of the outline. Cut between the holes. Remember, with hand saws, have the good surface up; with a saber saw, face down.

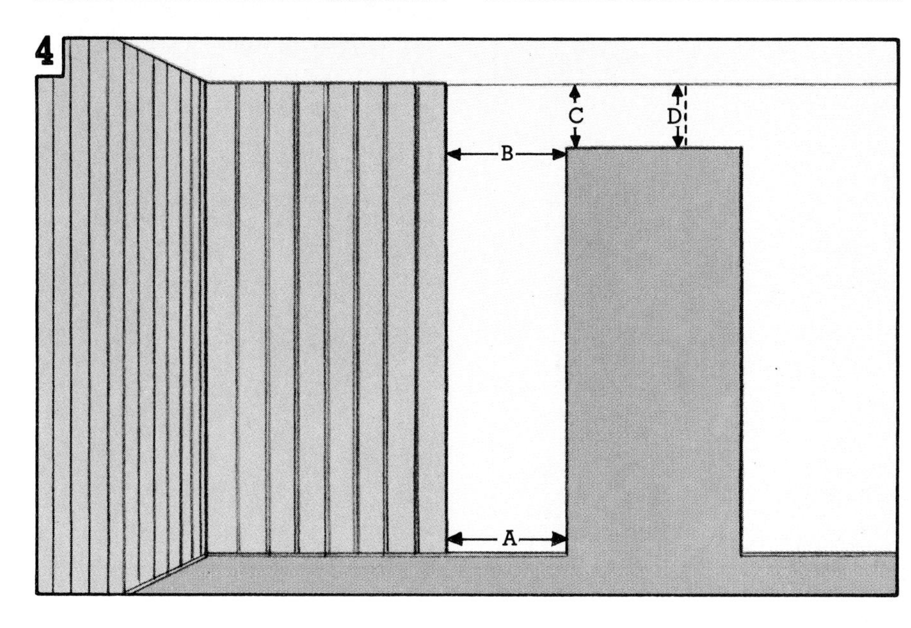

4 If you need to panel around door or window openings, again measure over from the last panel installed. Measure up from the floor or down from the ceiling to find the height of the opening. Notice the A, B, C, D sequence of measurements here. If possible, always try to lay out panels so their seams fall over the center of openings.

Paneling Around Corners and Obstructions *(cont.)*

5 To negotiate around a brick or stone fireplace (or other similar obstructions), measure the distance from the last installed panel to the farthest point along the irregular surface. Measure the mantel's height, too, allowing for the ¼-inch gap along the floor.

Transfer your findings to the panel, connect the cut lines with the help of a square or some other straightedge, and saw out the waste area. Now hold the panel plumb and alongside the irregular surface at the proper elevation. Temporarily nail the panel across the top. With the compass set at the width the panels overlap, scribe the contours of the irregular surface onto the panel, using a compass as shown here.

6 Following this, cut out the scrap piece with a coping saw (or a saber saw with a fine-tooth blade) and secure the panel in place.

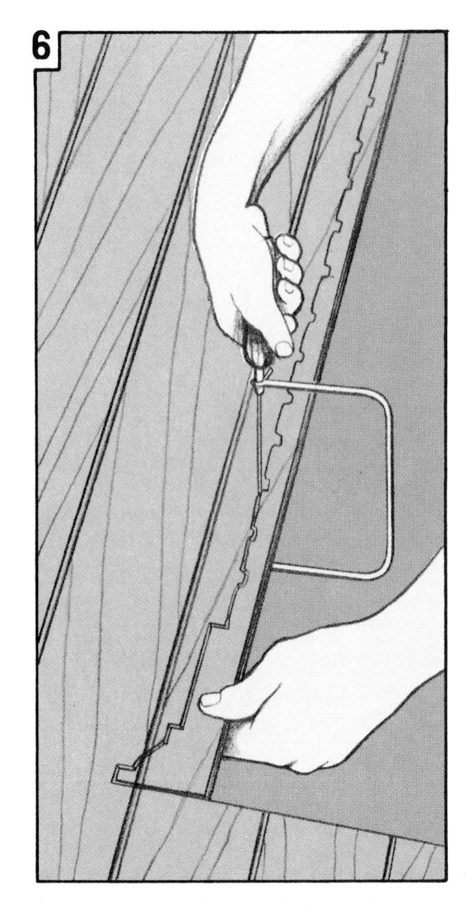

Finishing the Job

Adding the final touches to a paneling job involves two operations. The first—cutting precisely and nailing up decorative molding—allows you to cover the raw edges and gaps at the seams, floor, ceiling, and corner joints. **For information on what molding goes where, and how to order it, turn to page 20.**

The second operation—hiding nailheads and holes with color-matched putty—permits panels and molding to appear as blemish free as possible. Of course, if you've used color-matched nails, you needn't bother doing this.

1 As you can see here, you can cut lengths of baseboard molding in a variety of ways, depending on the situation and how you choose to deal with it. Along walls you can go with a 90° butt splice or—for a better job of hiding the seam—opt

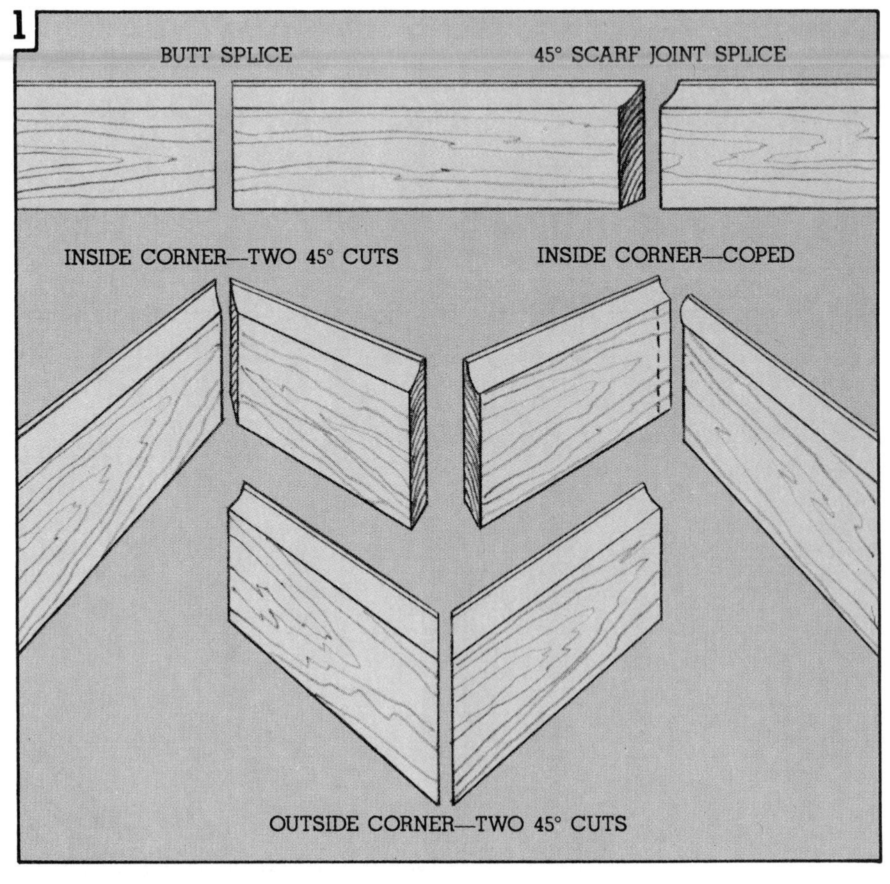

for a mitered 45° scarf joint splice. For inside corners, go with a mitered or coped joint—each is equally attractive. And for outside corners, always miter.

To accurately measure and mark individual pieces of molding, place untrimmed pieces where they're to be installed along the wall. Mark the cutoff point, then strike the needed angle, using a T bevel or combination square.

To cut angles on most moldings (except ceiling moldings) simply place the material in a miter box in the position it will be in when installed, adjust the saw to the desired angle, and make the cut. In some cases you may need to make a bevel or coped cut (see pages 36-37 on angle, bevel, and coped cuts).

Now secure the molding to the wall, driving finish nails at each 16-inch stud location. If the molding splits when you drive the nails, drill pilot holes first.

2 With cove or crown ceiling molding, the strategy is a little different. First, establish the top and bottom of the piece you're installing, identifying the surface that's to go against the ceiling. (Usually it's narrower than the surface that butts the wall.) Then look over the sketch to find out how to cut ceiling molding. Notice that in each case you make the 45-degree cuts while the molding rests upside down in the miter box.

3 Having placed the last strip of molding, go back and tap all nails 1/32 to 1/16 into the panels and molding using a nail set.

4 Follow this up by filling all nail holes and any scratches with a color-matched putty stick.

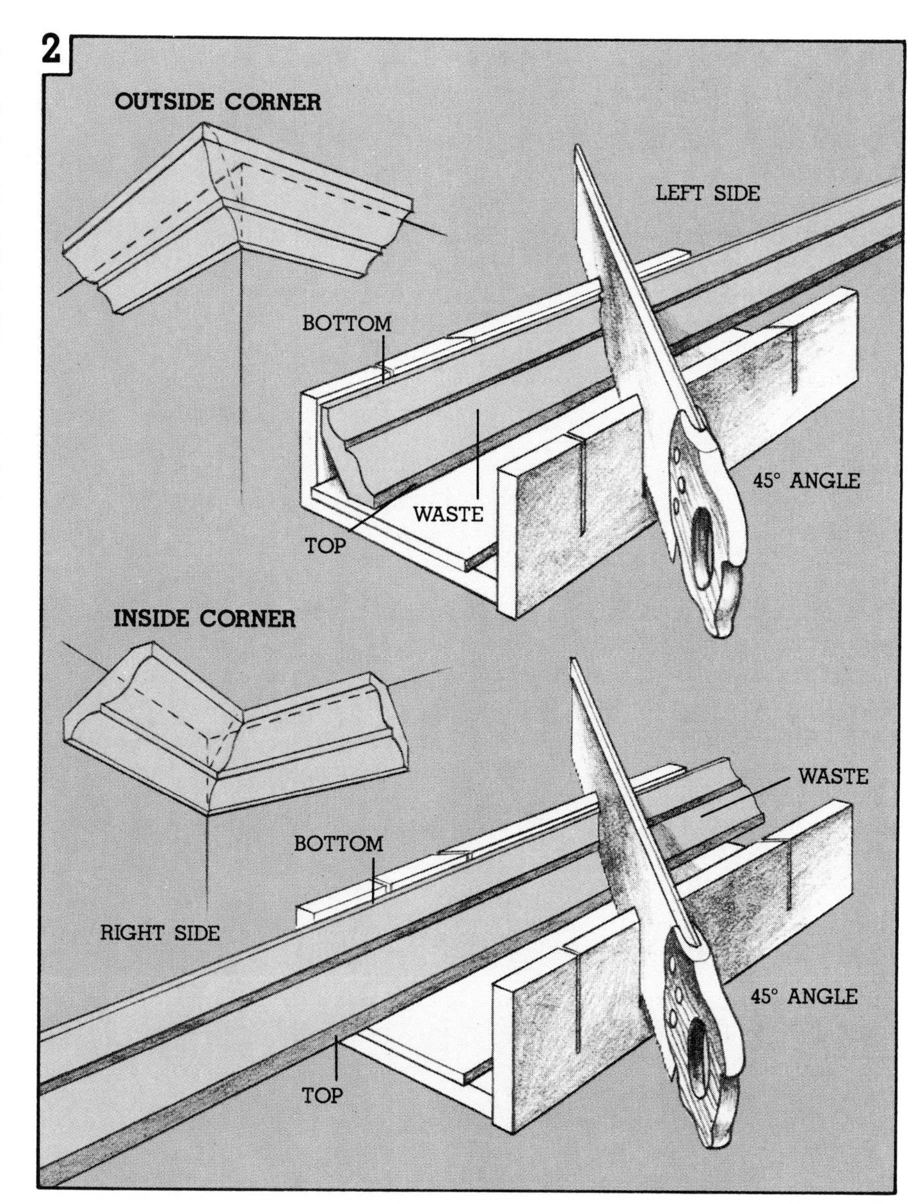

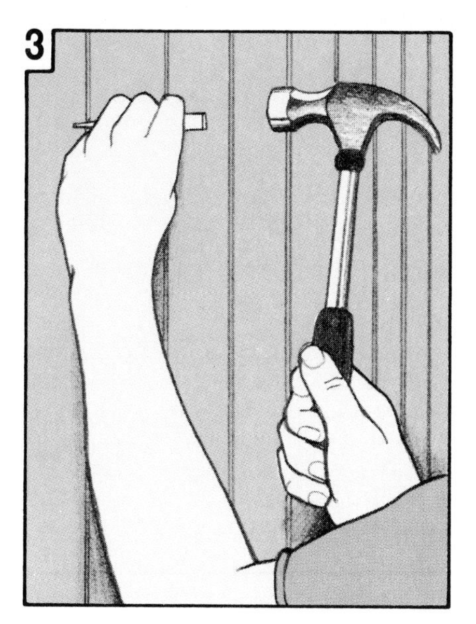

Glossary

Knowing the meaning of the entries on this page can make your next carpentry project easier and more fun. For terms not listed, or for more about those that are, refer to the index.

Actual dimension—The true size of a piece of lumber, after milling and drying. See also *nominal dimension.*

Bevel cut—An angle cut through the thickness of a piece of wood.

Board foot—The standard unit of measurement for wood. One board foot is equal to a piece 12×12×1 inches (nominal).

Board—Lumber that is less than 2 inches thick and over 3 inches wide.

Butt joint—The joint formed by two pieces of material when they meet *end to end* or *end to face or edge.*

Clinch—To hammer the exposed tip of a nail at an angle, bending its point into the surrounding wood for added joint strength.

Coped cut—A profile cut made in the face of a piece of molding that allows for butting it against another piece at an inside corner.

Counterbore—To drive a screw below the surface of the surrounding wood. The void created is filled later with putty, or plugged.

Countersink—To drive the head of a nail or screw so its top is flush with the surface of the surrounding wood.

Dado joint—A joint formed when the end of one member fits into a groove cut partway through the face of another.

Dimension lumber — Lumber that is 2 inches thick and at least 2 inches wide.

Filler—A pastelike compound used to hide surface imperfections in wood. Another type—pore filler—levels the surface of wood having a coarse grain.

Furring—Lightweight strips of wood applied to walls to provide a plumb nailing surface for paneling or drywall.

Gusset—A piece of wood nailed or screwed over a joint for added strength.

Header—The framing component spanning a door or window opening in a wall. A header supports the weight above it and serves as a nailing surface for the door or window frame.

Jamb—The top and side frames of a door or window opening.

Kerf—The void created by the blade of a saw as it cuts through a material.

Lap joint—The joint formed when one member overlaps another.

Ledger—A horizontal strip (quite often wood) that's used to provide support for the ends or edges of other members.

Level—The condition that exists when any type of surface is at true horizontal.

Lineal foot—The actual length of a board or piece of molding. See also *board foot.*

Miter joint—The joint formed when two members that have been cut at the same angle (usually 45°) meet.

Nominal dimension—The stated size of a piece of lumber, such as a 2×4 or a 1×12. The *actual dimension* is somewhat smaller.

On-center (OC)—A phrase used to designate the distance from the center of one regularly spaced framing member to the center of the next.

Pilot hole—A small hole drilled into a wooden member to avoid splitting the wood when driving a screw or nail.

Plumb—The condition that exists when a member is at true vertical.

Pressure-treated wood—Lumber and sheet goods impregnated with one of several solutions to make the wood virtually impervious to moisture and weather.

Rip—To saw lumber or sheet goods parallel to its grain pattern.

Roughing-in—The framing stage of a carpentry project. This framework later is concealed in the finishing stages.

Sealer—A protective coating (usually clear) applied to wood and metal.

Shim—A thin piece of wood or other material used to fill a gap between two adjoining components or to help establish level or plumb.

Square—The condition that exists when one surface is at a 90-degree angle to another.

Veneer—A thin layer of decorative wood laminated to the surface of a more common wood.

Warp—Any of several lumber defects caused by uneven shrinkage of wood cells. (See page 14 for several examples.)

Index

A-B

C

D-G

H-L

M-O

P-R

S

T-Z